T0271822

Neural Networks, Machine Learning, and Image Processing

This textbook comprehensively discusses the latest mathematical modeling techniques and their applications in various areas such as fuzzy modeling, signal processing, neural network, machine learning, image processing, and their numerical analysis. It further covers image processing techniques such as the Viola-Jones method for face detection and fuzzy approach for person video emotion. It will serve as an ideal reference text for graduate students and academic researchers in the fields of mechanical engineering, electronics, communication engineering, computer engineering, and mathematics.

Features

- Discusses applications of neural networks, machine learning, image processing, and mathematical modeling
- Provides simulation techniques in machine learning and image processing-based problems
- Highlights artificial intelligence and machine learning techniques in the detection of diseases
- Introduces mathematical modeling techniques such as wavelet transform, modeling using differential equations, and numerical techniques for multi-dimensional data
- Includes real-life problems for better understanding

This book presents mathematical modeling techniques such as wavelet transform, differential equations, and numerical techniques for multi-dimensional data. It will serve as an ideal reference text for graduate students and academic researchers in diverse engineering fields such as mechanical, electronics and communication, and computer engineering.

Neural Networks, Machine Learning, and Image Processing
Mathematical Modeling and Applications

Edited by
Manoj Sahni, Ritu Sahni, and
José M. Merigó Lindahl

CRC Press
Taylor & Francis Group
Boca Raton London New York

CRC Press is an imprint of the
Taylor & Francis Group, an **informa** business

First edition published 2022
by CRC Press
6000 Broken Sound Parkway NW, Suite 300, Boca Raton, FL 33487-2742

and by CRC Press
4 Park Square, Milton Park, Abingdon, Oxon, OX14 4RN

CRC Press is an imprint of Taylor & Francis Group, LLC

Library of Congress Cataloging-in-Publication Data
Names: Sahni, Manoj, editor. | Sahni, Ritu, editor. | Merigó Lindahl, José M., editor.
Title: Neural networks, machine learning, and image processing : mathematical modeling and applications / edited Manoj Sahni, Ritu Sahni, José M. Merigó Lindahl.
Description: First edition. | Boca Raton : CRC Press, 2023. |
Includes bibliographical references and index.
Identifiers: LCCN 2022031271 (print) | LCCN 2022031272 (ebook) |
ISBN 9781032300146 (hbk) | ISBN 9781032300160 (pbk) | ISBN 9781003303053 (ebk)
Subjects: LCSH: Neural networks (Computer science) | Machine learning. |
Image processing–Digital techniques.
Classification: LCC QA76.87 .N4875 2023 (print) | LCC QA76.87 (ebook) |
DDC 006.3/2–dc23/eng/20220816
LC record available at https://lccn.loc.gov/2022031271
LC ebook record available at https://lccn.loc.gov/2022031272

ISBN: 9781032300146 (hbk)
ISBN: 9781032300160 (pbk)
ISBN: 9781003303053 (ebk)

DOI: 10.1201/9781003303053

Typeset in Sabon
by codeMantra

Contents

Preface

Mathematical modeling is a field that provides fresh insights into natural phenomena by approximating and formulating physical situations. Scientists gather real-world data relevant to a specific topic through observations or experiments and then develop mathematical models to explain and predict the behavior of the real-world object whose scientific model they created. These models are close representations of real objects, not exact replicas. Thus, it is essential to work on the development of more precise models by using various mathematical tools. Mathematical modeling becomes easier with the help of machine learning tools and neural network algorithms. Neural network algorithms, in fact, work in the same way that our brains do. We begin by observing any real-life phenomenon with our eyes or collecting data with machines such as microscopes, telescopes, and cameras, and then we process that data by hypothesizing the underlying principles hidden in the phenomenon. We perform more and more experiments for further verification and then provide the results. The neural network also receives inputs in the form of numerical data, text, images, or any type of pattern, then processes the inputs by translating those data through various algorithms, and finally generates outputs.

This book aims to provide the most recent research on the development of various mathematical techniques in the area of neural networks, as well as the use of various machine learning techniques for better natural science modeling. It contains predictive models related to day-to-day problems, biological problems, engineering problems, and many other advancements in mathematical techniques. The aim is that models may be able to provide a more precise view, or at the very least a better understanding, of a real object or system. This book is divided into two parts, viz. Mathematical Modeling and Neural Network's Mathematical Essence and Simulations in Machine Learning and Image Processing.

The first section is mainly based on some mathematical modeling in the medical area, such as modeling of sarcopenia disease and modeling based on COVID-19 pandemic; modeling based on engineering problems related to automotive applications; monitoring of cylindrical bearing; and other decision-making problems such as university course scheduling for uncertainly generated courses and deep learning approach to detect text-based hate speech. The second section is related to machine learning and image processing topics such as problems based on data security, machine accuracy, detection of disease using machine learning, and problems based on image processing.

In this way, this book is very important for students, researchers, engineers, and computer scientists as it contains various applications of mathematical modeling containing strategies for the solutions of the problems as well as a systematic understanding of the modeling of any real-life problems and also the model of the latest system and technologies. It also provides techniques for the efficient use of latest computerized mathematical techniques for the betterment of the world. This book not only contains real-life problems, but also provides precise theory related to the latest research on mathematical modeling, machine learning, and numerical techniques in an uncertain environment.

We would like to express our sincere gratitude to all of the authors for their contributions in the form of chapters, and we hope that all readers will benefit from this book and be successful as a result of the tremendous effort put forth in it.

MATLAB® is a registered trademark of The MathWorks, Inc. For product information, please contact:

The MathWorks, Inc.
3 Apple Hill Drive
Natick, MA 01760-2098 USA
Tel: 508-647-7000
Fax: 508-647-7001
E-mail: info@mathworks.com
Web: www.mathworks.com

Editors

Dr. **Manoj Sahni** is a dedicated and experienced mathematics teacher and researcher with more than seventeen and a half years of experience and currently serving as an Associate Professor and Head in the Department of Mathematics, School of Technology, Pandit Deendayal Energy University, Gandhinagar, Gujarat, India. He has an excellent academic background with M.Sc. (Mathematics with specialization in Computer Applications), from Dayalbagh Educational Institute (Deemed University), Agra, M. Phil. from IIT Roorkee, and a Ph.D. in Mathematics from Jaypee Institute of Information Technology (Deemed University), Noida, India. He has published more than 70 research papers in peer-reviewed journals (SCI/SCIE/ESCI/Scopus), conference proceedings (Scopus indexed), and book chapters (Scopus indexed) and with reputed publishers such as Springer and Elsevier. He also serves as an advisory board member, technical committee member, and reviewer for many international journals of repute and conferences. He has organized 1st, 2nd, and 3rd International Conferences on Mathematical Modeling, Computational Intelligence Techniques, and Renewable Energy (MMCITRE) held on February 21–23, 2020, February 06–08, 2021, and March 04–06, 2022, respectively. He participated in the scientific committee of several international conferences and associations and also delivered many expert talks at national and international levels. He has organized many seminars, workshops, and short-term training programs at Pandit Deendayal Energy University (PDEU), and various other universities. He has also organized Special Symposia in an International Conference (AMACS2018) on Fuzzy Set Theory: New Developments and Applications to Real-Life Problems held in London in 2018. In addition, he is a member of many international professional societies, including the American Mathematical Society (AMS), Society for Industrial and Applied Mathematics (SIAM), IEEE, Mathematical Association of America (MAA), Forum for Interdisciplinary Mathematics (FIM), Indian Mathematical Society (IMS), and IAENG.

Dr. Ritu Sahni is a dedicated and experienced mathematics teacher and researcher with more than 15 years of experience who is actively involved in both undergraduate and postgraduate levels of teaching students. She has a strong academic background by completing B.Sc. (Mathematics), M. Sc. (Mathematics with specialization in Computer Applications) at Dayalbagh Educational Institute, Agra, U.P., India, and Ph.D. at Jaypee Institute of Information Technology (JIIT), Noida, India. During her Ph.D., she also served her institution (JIIT, Noida, India) by teaching undergraduate and postgraduate students, and after the completion of her doctorate degree, she served various institutions including Navrachana University, Vadodara, Gujarat, India, and Institute of Advanced Research (IAR), Gandhinagar, India, as an Assistant Professor, and presently, she is working as a Visiting Faculty at Pandit Deendayal Energy University, Gandhinagar, Gujarat, India. She worked with her students on various research-related problems and published their work in reputed journals (SCI/SCIE/ESCI/Scopus) such as Elsevier and Springer. She has published more than 45 research papers in peer-reviewed international journals and conferences. She is the reviewer of many international journals and is a member of the Indian Science Congress and many other well-renowned societies. She has also organized seminars and workshops and also coordinated the 2nd and 3rd International Conferences on Mathematical Modelling, Computational Intelligence Techniques, and Renewable Energy (MMCITRE) held on February 06–08, 2021, and March 04–06, 2022. She has delivered talks and presented many research articles in various international conferences in India and abroad. Her research fields include fixed point theory and its applications, numerical methods, fuzzy decision-making systems, and other allied areas. She is a devoted faculty member, committed researcher, and enthusiast striving to improve the institution's educational offerings and research fields for societal growth.

Prof. José M. Merigó Lindahl is a Professor at the School of Information, Systems & Modelling at the Faculty of Engineering and Information Technology at the University of Technology Sydney (Australia) and Part-Time Full Professor at the Department of Management Control and Information Systems at the School of Economics and Business at the University of Chile. Previously, he was a Senior Research Fellow at the Manchester Business School, University of Manchester (the UK), and an Assistant Professor at the Department of Business Administration, University of Barcelona (Spain). He holds a master's and a Ph.D. degree in Business Administration from the University of Barcelona. He also holds a B.Sc. and a M.Sc. degree from Lund University (Sweden). He has published more than 500 articles in journals, books, and conference proceedings, including journals such as Information Sciences, IEEE Computational Intelligence Magazine, IEEE Transactions on Fuzzy Systems, European

Journal of Operational Research, Expert Systems with Applications, International Journal of Intelligent Systems, Applied Soft Computing, Computers & Industrial Engineering, and Knowledge-Based Systems. He has also published several books with Springer and with World Scientific. He is on the editorial board of several journals, including *Computers & Industrial Engineering*, Applied Soft Computing, Technological and Economic Development of Economy, *Journal of Intelligent & Fuzzy Systems*, International Journal of Fuzzy Systems, Kybernetes, and Economic Computation and Economic Cybernetics Studies and Research. He has also been a guest editor for several international journals, member of the scientific committee of several conferences, and reviewer in a wide range of international journals. Recently, Thomson & Reuters (Clarivate Analytics) has distinguished him as a Highly Cited Researcher in Computer Science (2015–present). He is currently interested in decision making, aggregation operators, computational intelligence, bibliometrics, and applications in business and economics.

Contributors

Saraswati Acharya
Department of Mathematics
Kathmandu University
Dhulikhel, Nepal

Mustafa Africawala
Department of Computer Science and Engineering
Pandit Deendayal Energy University
Gandhinagar, India

Mohammed Ghazi Al-Khaiyat
Department of Software Engineering
Jadara University
Irbid, Jordan

Tareq Al-shaikh
Department of Software Engineering
Jadara University
Irbid, Jordan

Sunil B. Bhoi
Department of Applied Mathematics and Humanities
Sardar Vallabhbhai National Institute of Technology
Surat, India

V. Rajkumar Dare
Madras Christian College
Chennai, India

Pallavi Dash
Electrical Engineering Department
International Institute of Information Technology
Bhubaneswar, India

Nisarg Dave
Department of Computer Science and Engineering
Pandit Deendayal Energy University
Gandhinagar, India

Kalyani Desikan
Department of Mathematics
Vellore Institute of Technology
Chennai, India

Namrata Devata
Electrical Engineering Department
International Institute of Information Technology
Bhubaneswar, India

Harsh S. Dhiman
Electrical Engineering Department, Adani Institute of Infrastructure
   Engineering,
Ahmedabad, India

Jayesh M. Dhodiya
Department of Applied Mathematics and Humanities Sardar Vallabhbhai
   National Institute of Technology
Surat, India

R. Telang Gode
Department of Mathematics
National Defence Academy, Pune, India

Devang Goswami
Electrical Engineering Department, Adani Institute of Infrastructure
   Engineering,
Ahmedabad, India

Dil Bahadur Gurung
Department of mathematics
Kathmandu University
Dhulikhel, Nepal

Abram Magdy Jonan
Department of Computer Science
Jadara University
Irbid, Jordan

Bhaumik Joshi
Electrical Engineering Department, Adani Institute of Infrastructure
   Engineering
Ahmedabad, India

Nisarg Kapkar
Department of Computer Science and Engineering
Pandit Deendayal Energy University
Gandhinagar, India

Debani Prasad Mishra
Electrical Engineering Department
International Institute of Information Technology
Bhubaneswar, India

N. Mohana
Department of Mathematics
Vellore Institute of Technology
Chennai, India

Shaker Mrayyen
Department of Computer Science
Jadara University
Irbid, Jordan

S.V.S.S.N.V.G Krishna Murthy
Department of Applied Mathematics
D.I.A.T
Pune, India

M. Palav
Department of Applied Mathematics
D.I.A.T
Pune, India

Divyang H. Pandya
Department of Mechanical Engineering
LDRP Institute of Technology and Research
Gandhinagar, India

Meghna Parikh
Department of Mathematics
Pandit Deendayal Energy University
Gandhinagar, India

Umang Parmar
Department of Mechanical Engineering
LDRP Institute of Technology and Research
Gandhinagar, India

Bansi Patel
Computer Science Department
Pandit Deendayal Energy University
Gandhinagar, India

Samir Patel
Computer Science Department
Pandit Deendayal Energy University
Gandhinagar, India

Sanyam Raina
Department of Computer Science and Engineering
Pandit Deendayal Energy University
Gandhinagar, India

M. Ramasubramanian
Chemical Engineering Department
Coimbatore Institute of Technology
Coimbatore, India

Saketram Rath
Department of Electrical Engineering
International Institute of Information Technology
Bhubaneswar, India

DhruvinSinh Rathod
Computer Science Department
Pandit Deendayal Energy University
Gandhinagar, India

Manoj Sahni
Department of Mathematics
Pandit Deendayal Energy University
Gandhinagar, India

Ritu Sahni
Department of Mathematics
Pandit Deendayal Energy University
Gandhinagar, India

Jahnavi Shah
Department of Computer Science and Engineering
Pandit Deendayal Energy University
Gandhinagar, India

Dev Chandra Shrestha
Department of mathematics
Kathmandu University
Dhulikhel, Nepal

M. Thirumarimurugan
Chemical Engineering Department
Coimbatore Institute of Technology
Coimbatore, India

Arwa Zabian
Department of Software Engineering
Jadara University
Irbid, Jordan

Mathematical modeling and neural network's mathematical essence

Chapter 1

Mathematical modeling on thermoregulation in sarcopenia

Dev Chandra Shrestha, Saraswati Acharya,
and Dil Bahadur Gurung
Kathmandu University

CONTENTS

1.1 INTRODUCTION

Sarcopenia is a type of age-related loss of skeletal muscle mass and strength that occurs in people approximately after the age of 50 years. The age-related loss of muscle mass, muscle strength, and physical performance is the key factor to determine sarcopenia. It also happens in chronic diseases, occurs with changes in lifestyle and habits, and decreases the metabolic rate.

Sarcopenia is a universal disease that leads to a reduction in the quality of life and rises in disability prevalence with aging. Muscle strength losses occur dramatically after 70 years of age. At the beginning of life, muscle mass increases, but from the age of 50, it decreases approximately 8% every decade [1]. An eighty-year-old person loses 30% more muscle mass than

DOI: 10.1201/9781003303053-2

3

what a 70-year-old person does [2]. The impact of muscle mass decrease on total body mass may be responsible for the decrease in the rate of intake of food energy [3]. The basal metabolic rate (BMR) slows down and affects the body temperature due to a decrease in the consumed food energy. A human above 65 years of age takes only 96 kJ/day energy from nutrition [2]. Due to the reduction in the metabolism, the body core temperature slightly falls with aging [3, 4].

Muscle mass plays a crucial role in physical fitness. On building muscles, the body burns energy and fat all the time. In sarcopenia, fat increases, while fat-free mass (FFM) decreases. Having excess fat increases the risk of developing coronary artery disease, diabetes, hyperlipidemia, and hypertension [5]. The BMR decreases with increasing age of people, showing elders have less BMR than adults. Due to less BMR, hormonal glands secrete fewer hormones to regulate the slow metabolic process. Low muscle strength associates with an increase in metabolic syndrome, type II diabetes, and some mortality cases. Research has shown that strength and resistance training can reduce or prevent muscle loss. In lean muscle tissue, energy burns. The average energy expenditure of persons in the age range of 50–60 years is 1,679.44 kcal/day, and it is 1,407.15 kcal/day in 80–90 years. The estimated average energy expenditure in persons aged 50–90 years is presented in Table 1.1.

The rate of loss of muscle mass varies and depends on age, gender, and type of physical exercise. Building muscles affects body weight as well. In this process, the body gains muscle mass, loses fat, and increases body weight. The level of sarcopenia is determined by the primary amount of muscle mass and the speed at which muscle mass declines. The stage of sarcopenia increases due to decreased nutrient intake, low physical activity, and the appearance of chronic disease. The protection of muscle mass and prevention of sarcopenia can help to prevent the decrease in BMR although body weight increases with aging. In aging, body weight decline in older adults is higher than in younger adults. The body weight is lowest after the age of 80 years and above. The weight loss in older adults is caused by various reasons such as hormonal changes, deficiency in protein consumption, changes in protein metabolism, and decrease in the muscular mass, which may lead to morbidity and physical inability. The average body weight of persons in the age range 50–90 years is presented in Table 1.2.

Table 1.1 An estimation of the energy expenditure associated with old age [6]

Age (years)	Energy expenditure (kcal/day)
50–60	1,679.44
60–70	1,599.23
70–80	1,485.42
80–90	1,407.15

Table 1.2 Bodyweight of different ages [7]

Age	Body weight (kg)
50–60	77.85
60–70	77.49
70–80	73.83
80–90	74.42

Metabolism and skin thickness also play a vital role in the thermoregulation of human body in sarcopenia persons. Muscle mass is the largest component; around 40% of the whole-body weight contributes only 20% of BMR. The brain and liver make up about 4% of body weight, contributing to 40%–45% of BMR, while the heart, kidneys, and gut contribute about 35%–40% BMR of the whole body [8]. This shown the major organs such as brain, liver, kidney, heart, and gut represent approximately 80% of BMR and play a major role in the BMR of the whole body. The BMR depends on the body weight and can be calculated if the body weight is known [6]. But due to aging, the BMR, fat-free mass (FFM), and physical activities are decreased. So, the body core temperature falls and it feels cold with aging due to metabolic effects.

The skin layers epidermis, dermis, and subcutaneous tissue (SST) region play an important role in the heat energy exchange from the body to the environment and environment to the body. Heat exchange between the skin layers depends on the thickness of the skin, its thermal conductivity, its location, and its heat transfer coefficient. The process is also influenced by the environment, age, and the presence of active sweat glands in the dermis. Normally, the thickness of the SST region varies, and its temperature changes notably toward the outer skin surface from the body core with the change in environmental temperature. This is because of heat loss from the skin surface and deposited fat in the subcutaneous tissue [9]. After the age of 50 years, the temperature of the epidermis and dermis layers increases due to the decrease in their thickness, but the body core temperature decreases due to the decrease in the BMR, which provides the body with thermoregulation by the body mechanism [10].

Researchers studied the effect of blood flow on heat transfer in living tissues and developed bio-heat transfer models. In 1948, abio-heat transfer model was first developed by Harry Pennes' for predicting heat transfer in the human forearm. This model describes the blood perfusion effect and metabolic rate that occurs in tissue, which are incorporated in the standard thermal diffusion equation given by [11]:

$$\rho_t c_t \frac{\partial \theta}{\partial t} = \nabla \cdot (K \nabla \theta) + \rho_b c_b \omega_b (\theta_a - \theta) + M_A(t)$$

where

ρ_t: Tissue density (kg/m^3),
c_t: Tissue specific heat capacity (J/kg°C),
K:Tissue thermal conductivity (w/m.°C),
θ: Tissue temperature (°C),
ω_b : Blood per fusion rate (/s),
θ_a: Artery temperature (°C), and
$M_A(t)$: Metabolic heat generation rate (w/m^3).

Waalen and Buxbaum [12] compared the body temperatures of men and women and found that women have a slightly higher body temperature than men and the mean temperature decreases by 0.17°C in the age group of 70–80 years. The effect of blood flow on tissue temperature was explained by Coccarelli et al. [13]. Researchers proposed that the change in arterial blood flow does not affect the distribution of tissue temperature significantly with aging. Kenney and Munce [4] explained the thermoregulation in young adults and the elderly due to cold stress. They investigated that peripheral vasoconstriction is reduced and metabolic heat production decreased in the elderly when compared to young adults. Bianchetti et al. [1] studied the reduction process of sarcopenia with aging. They recommended that resistance training and amino acid supplementation are keys to it. Keil et al. [14] determined the relationship between body temperature and aging. They investigated that women have a higher body temperature, yet live longer than men. Henry [8] developed a model relating the BMR, age, and bodyweight based on their numerical values. Furthermore, he provided an approximation of the BMR values of different organs.

Previously, researchers have not studied the mathematical model of temperature distribution in three layers of skin with aging. So, this model proposes to estimate the BMR and thickness of skin layers and to investigate the temperature of the epidermis, dermis, and subcutaneous tissue with aging. The finite element method is used to discretize the domain of skin layers and obtain the sensible temperature profiles of the dermal part.

1.2 DISCRETIZATION

Skin is an important organ of the body and has numerous functions. It forms a physical barrier to the environment to help in regulating and maintaining the body temperature. The thickness of the skin layer is important for maintaining body temperature normal. In this chapter, each layer distance is measured perpendicularly from the outer skin surface toward the body core. Let L_1, L_2, and L_3 be the distance of epidermis, dermis, and subcutaneous tissue, respectively, measured from the outer surface of the skin toward the body core. This provides that the thickness of the epidermis is

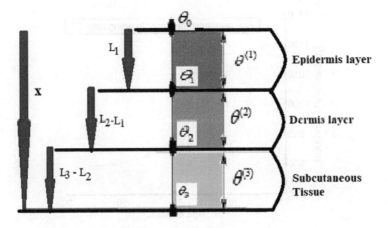

Figure 1.1 Schematic diagram of the three layers of skin.

L_1–L_0 the dermis is L_2–L_1, and the subcutaneous tissue is L_3–L_2. The schematic diagram of the skin layer thickness of the human body is presented in Figure 1.1.

Let θ_0, θ_1, and θ_2 be the nodal temperatures at a distance of $x=0$, $x=L_1$, and $x=L_2$, respectively. θ_3 is the body core temperature at a distance of $x=L_3$. Let $\theta^{(0)}$, $\theta^{(1)}$, and $\theta^{(2)}$ be the temperature functions of the epidermis, dermis, and subcutaneous tissue, respectively.

1.3 MODELING AND SIMULATION OF BASAL METABOLIC RATE AND SKIN LAYER THICKNESS

1.3.1 Basal metabolic rate model

The BMR of a healthy adult is 1,114 w/m³, and its value is influenced by factors such as body size, age, hormones, gender, and body weight [15]. The BMR of a birth child is 2,183.04 w/m³ and has a maximum value in the first 3 months of life [8]. Afterward, the BMR declines with age. The metabolism is low after the age of 50 years and behaves as a decreasing logistic phenomenon due to the loss of muscle mass, as shown in Figure 1.2. At the age of 50–60 years, the average BMR is approximately 1,022.8 w/m³, and at the age of 80–90 years, the approximate BMR value is 895.58 w/m³. The approximation BMR at different ages is presented in Table 1.3. The BMR equation behavior $M_A(t)$ with aging is consider as [15]:

$$M_A(t) = S_m + \frac{(M_c - S_m)}{1 + e^{-\omega(t - t_{sg})}} \tag{1.1}$$

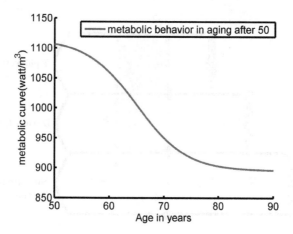

Figure 1.2 Metabolic rate behavior with aging.

Table 1.3 The basal metabolic rate estimation
with aging

Age (years)	Basal metabolic rate (w/m³)
50–60	1,022.8
60–70	978.45
70–80	948.83
80–90	895.56

where

S_m: BMR of normal human body,

t : Age in years,

M_c : BMR of aging,

ω : BMR controlled parameter, and

t_{sg} : Sigmoid's mid-point of the curve.

1.3.2 Skin layer thickness model

The thickness of the skin increases over the first two decades of life, even though the number of cell layers remains the same. The dermis layer thickness decreases by 2% per year, while the epidermis layer thickness decreases by 6.4% every decade. The subcutaneous tissues lose fatty cushion, while the basement membrane thickness increases and the total thickness remains almost constant with age [16].

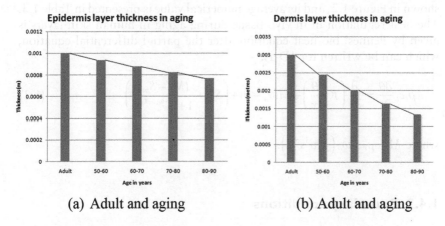

(a) Adult and aging   (b) Adult and aging

Figure 1.3 Changes in skin layers' thicknesses with aging.

Table 1.4 Calculation of skin layers' thickness with aging

Age (years)	Epidermis layer thickness $((L_1)\ m)$	Dermis layer thickness $((L_2-L_1)\ m)$	Subcutaneous tissue thickness $((L_3-L_2)\ m)$
50–60	9.36×10^{-4}	24.51×10^{-4}	50.00×10^{-4}
60–70	8.76×10^{-4}	20.03×10^{-4}	50.00×10^{-4}
70–80	8.20×10^{-4}	16.37×10^{-4}	50.00×10^{-4}
80–90	7.68×10^{-4}	13.38×10^{-4}	50.00×10^{-4}

In general, the thickness of epidermis, dermis, and subcutaneous tissue is taken as 0.001, 0.003, and 0.005 m, respectively, in a healthy adult [17]. In the age range of 50–60 years, the epidermis layer thickness becomes 9.36×10^{-4}m and the dermis layer thickness becomes 24.51×10^{-4}m. In the age range of 80–90 years, the epidermis and dermis layers' thickness reduces to $7.68 \times 10^{-4}$and 13.38×10^{-4}m, respectively. In this model, we assume the thickness of the epidermis, dermis, and subcutaneous tissue to be L_1, L_2-L_1, and L_3-L_2, respectively. The reduction in the epidermis and dermis layers' thickness is graphically shown in Figure 1.3a and b, and it's numerical values with age are presented in Table 1.4.

1.4 MATHEMATICAL MODEL AND BOUNDARY CONDITIONS

1.4.1 Mathematical model

The metabolic rate $M_A(t)$ slows down with increasing age and falls below the average BMR after 50 years due to the loss of muscle mass caused by the energy expenditure rate. The decreasing behavior of BMR with aging is shown in Figure 1.2, and its average numerical value is presented in Table 1.3. The heat regulation in in vivo tissue during aging in human skin layers is given by Pennes' bio-heat equation after the partial differential equation, which can be written for 1D as:

$$\rho_t c_t \frac{\partial \theta}{\partial t} = \frac{\partial}{\partial x}\left(K \frac{\partial \theta}{\partial x}\right) + M(\theta_a - \theta) + \left(S_m + \frac{(M_c - S_m)}{1 + e^{-\omega(t-t_{sg})}}\right) \tag{1.2}$$

where $M = \rho_b c_b \omega_b \left(\text{J/m}^3\text{s}°\text{C}\right)$.

1.4.2 Boundary conditions

1.4.2.1 Boundary condition at x = 0 (skin surface)

The skin is exposed to the atmosphere, and the heat flux is dissipated at the outer surface $x=0$. Evaporation, convection, and radiation are the processes that allow heat to escape from the outer surface. So, the net heat flux calculated by mixed boundary condition changes to

$$K \frac{\partial \theta}{\partial x}\big|_{x=0} = h_a \left(\theta - \theta_\infty\right) + L_a E \tag{1.3}$$

where $\frac{\partial}{\partial x}$ is the temperature derivative along outward normal to the boundary surface and $h_a = h_c + h_r$, where, h_c and h_r are the heat transfer coefficient due to convection and radiation heat flow, respectively. θ_∞, L_a, and E are the ambient temperature, latent heat capacity, and sweat evaporation rate of aging, respectively.

1.4.2.2 Boundary condition at x = L₃ (body core)

Aging slows down the movement of muscle mass, but all the metabolic heat energy dissipates instantaneously at that time. The body core temperature stays at 37°C. So, the Dirichlet' sinner boundary condition is assumed as:

$$\theta\left(L_3\right) = \theta_b = 37°\text{C}$$

where

θ_b: Temperature of the body core and
L_3: Total thickness of the skin.

1.5 SOLUTION OF THE MODEL

Using the finite element method, equation (1.2) can be solved using the boundary and initial conditions. For steady-state temperature, the variational integral form equation is given by

$$I[\theta(x,t)] = \frac{1}{2}\int_{\Omega}\left[K\left(\frac{d\theta}{dx}\right)^2 + M(\theta_a - \theta)^2 - 2\left(S_m + \frac{(M_c - S_m)}{1 + e^{\omega(t - t_{sg})}}\right)\theta \right]dx$$

$$+\frac{1}{2}h_a(\theta_0 - \theta_\infty)^2 + L_a E \theta_0$$

(1.4)

where

Ω: skin layered domain and
θ_0: skin surface temperature.

The various physiological and physical parameters associated with the epidermis, dermis, and subcutaneous tissue are mentioned in this model, as given in Table 1.5.

Let I_k, for $k = 1,2,3$, be the integral solutions of the three layers epidermis, dermis and subcutaneous tissue, respectively, with $I = \sum_{k=1}^{3} I_k$. Solving the integrals I_1, I_2, and I_3 with parameters as mentioned in Table 1.5, we obtain the solutions I_1, I_2, and I_3 as functions of nodal values θ_0, θ_1, and θ_2 as given below:

$$I_1 = A_1 + B_1\theta_0 + D_1\theta_0^2 + E_1\theta_1^2 + F_1\theta_0\theta_1$$

(1.5)

$$I_2 = A_2 + B_2\theta_1 + C_2\theta_2 + D_2\theta_1^2 + E_2\theta_2^2 + F_2\theta_1\theta_2$$

(1.6)

$$I_3 = A_3 + B_3\theta_2 + C_3\theta_3 + D_3\theta_2^2 + E_3\theta_3^2 + F_3\theta_2\theta_3$$

(1.7)

where A_j, B_j, D_j, E_j, F_j, and C_z with $1 \leq j \leq 3$ and $2 \leq z \leq 3$ are all constants whose values confide in physical and physiological parameters of skin layers as given in Table 1.6. Furthermore, to optimize I, differentiate the system of equations (1.5)–(1.7) with respect to θ_0, θ_1, and θ_2 and set $\dfrac{dI}{d\theta_k} = 0$, for $k = 0$, 1, 2. After simplification, we get the system of equations in matrix form:

$$R\theta = V$$

(1.8)

Table 1.5 Assumption of parameters in the model

Physical and physiological parameters	Epidermis layer $0 \leq xz \leq L_1$	Dermis layer $L_1 \leq x \leq L_2$	Subcutaneous tissue $L_2 \leq x \leq L_3$
K	K_1	K_2	K_3
M	$M_1 = 0$	M_2	M_3
θ_a	$\theta_a^{(1)} = 0$	$\theta_a^{(2)} = \theta_b$	$\theta_a^{(3)} = \theta_b$
M_A	$M_{A(1)} = 0$	$M_{A(2)} = S_m + \dfrac{M_c - S_m}{1 + e^{-\omega(t - t_{sg})}}$	$M_{A(3)} = 2\left[S_m + \dfrac{M_c - S_m}{1 + e^{-\omega(t - t_{sg})}}\right]$
$\theta^{(k)}$	$\theta^{(1)} = \theta_0 + \left(\dfrac{\theta_1 - \theta_0}{L_1}\right) x$	$\theta^{(2)} = \dfrac{L_2\theta_1 - L_1\theta_2}{L_2 - L_1} + \left(\dfrac{\theta_2 - \theta_1}{L_2 - L_1}\right) x$	$\theta^{(3)} = \dfrac{L_3\theta_2 - L_2\theta_3}{L_3 - L_2} + \left(\dfrac{\theta_3 - \theta_2}{L_3 - L_2}\right) x$

Table 1.6 The values of the parameters used in the model [17,15]

Parameter	L_a	K_1	K_2	K_3	h_a	$M_2 = M_3$	$\rho_1 = \rho_2 = \rho_3$	$c_1 = c_2 = c_3$
Value	2.42×10^6	0.209	0.314	0.418	6.27	1,254	1,050	3,469.4
Unit	J/kg	w/m°C	w/m°C	w/m°C	w/m²°C	w/m³°C	kg/m³	J/kg°C

where

$$R = \begin{pmatrix} 2D_1 & F_1 & 0 \\ F_1 & 2(D_2 + E_1) & F_2 \\ 0 & F_2 & 2(D_3 + E_2) \end{pmatrix}, \theta = \begin{pmatrix} 0 \\ 1 \\ 2 \end{pmatrix}, V = \begin{pmatrix} -B_1 \\ -B_2 \\ -C_2 - B_3 - F_{33} \end{pmatrix}$$

For the age range of 50–60 years at ambient temperature 25°C and evaporation rate 0 kg/m²s, the numerical elements of matrices R and V are obtained
as:

$$R = \begin{pmatrix} 6922 & -223.29 & 0 \\ -223.29 & 352.44 & -63.54 \\ 0 & -63.54 & 309 \end{pmatrix}, \text{ and } V = \begin{pmatrix} -156.75 \\ -58.11 \\ 101.88 \end{pmatrix}$$

The augmented matrix of the system of equation (1.8) is

$$Q = \begin{pmatrix} 6922 & -223.29 & 0 & -156.75 \\ -223.29 & 352.44 & -63.54 & -58.11 \\ 0 & -63.54 & 309 & 101.88 \end{pmatrix}$$

Since both the rank of the coefficient matrix R and the rank of the augmented matrix Q are equal to 3, the system of equations (1.8) is consistent.
In a similar manner, the system of equations of matrices are consistent for
different age groups at $E = 0$ kg/m²s and $E = 0.00004$ kg/m²s and at constant environment temperature.

1.6 NUMERICAL RESULTS AND DISCUSSION

The values of BMR for the age groups of 50–60, 60–70, 70–80, and 80–90 years are 1,022.8, 978.45, 948.83, and 895.56 w/m³, respectively, and are shown in Table 1.3. The values of physical and physiological parameters used for numerical simulation are shown in Table 1.6.

At normal atmospheric temperature, the initial skin temperature is considered 23°C So the tissue temperature increases from the skin surface toward the body core temperature.

1.6.1 Temperature results

The temperature profiles of epidermis, dermis, and subcutaneous tissue during aging and adult are presented in Figures 1.4–1.8 and Table 1.7 at ambient temperature θ_∞ = 25°C. Figure 1.4a and b delineate the temperature profiles of dermis layers during aging and adult at $E = 0.00004$ kg/m²s. These results show that the steady temperature of the dermis layer is more by 0.17°C in the age group of 50–60 years and more by 0.32°C in the age group of 60–70 than in adults. This is because a thinner layer conducts heat energy more than the thicker layer.

The temperature profiles of epidermis layer and subcutaneous tissue are presented in Figure 1.5a and b, respectively, at $E = 0$ kg/m²s. Figure 1.5a reveals that the temperature of the epidermis layer is higher by 0.22°C in the age group of 70–80 years and higher by 0.17°C in the age group of 60–70 years than in adults. The subcutaneous tissue temperature is lower by 0.10°C in the age group of 80–90 years and lower by 0.04°C in the age group of 60–70 years than in adults, as shown in Figure 1.5b. These results exhibit that the subcutaneous tissue steady temperature has less variation than the epidermis layer temperature in adults compared to the elderly. This is because the subcutaneous tissue has a higher fat layer in the elderly than in adults.

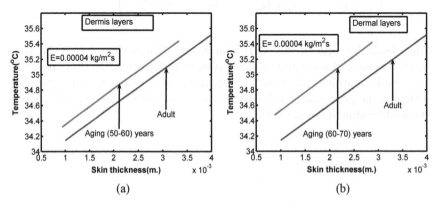

Figure 1.4   Observation of dermis layer temperatures at $E = 0.00004$ kg/m²s in adult with the elderly (a) (50–60) years and (b) (60–70) years.

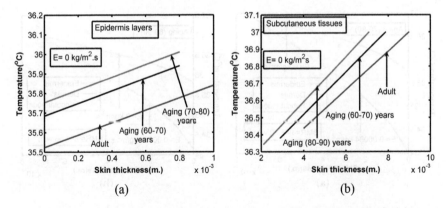

Figure 1.5 Comparison of epidermis layer and subcutaneous tissue temperatures at $E=0\,kg/m^2s$ in the elderly with age groups (a) (60–70), (70–80) years and adults, (b) (60–70), (80–90) years and adults.

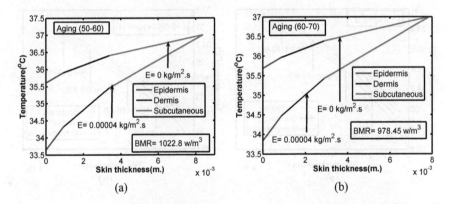

Figure 1.6 Observation of epidermis, dermis, and subcutaneous tissue temperatures at $E=0\,kg/m^2s$ and $E=0.00004\,kg/m^2s$ in the elderly with age (a) (50–60) years and (b) (60–70) years.

Figure 1.6a shows that the interface temperatures of epidermis, dermis, and subcutaneous tissue are, respectively, increased to 35.61°C, 35.90°C, and 36.40°C at $E = 0$ kg/m²s in the age group of 50–60 years. On increasing the evaporation rate from $E=0\,kg/m^2s$ to $E=0.00004\,kg/m^2s$, the epidermis, dermis, and subcutaneous tissue temperatures decreased by 2.97°C, 1.58°C, and 0.94°C, respectively, in the age group of 50–60 years. On observing Figure 1.6b, the epidermis, dermis, and subcutaneous tissue steady temperatures are higher by 1.58°C, 1.50°C, and 0.96°C at $E = 0$ kg/m²s than at $E = 0.00004$ kg/m²s in the age group of 60–70 years.

Figure 1.7a shows the epidermis, dermis, and subcutaneous tissue temperatures at $E = 0\,kg/m^2s$ and at $E = 0.00004\,kg/m^2s$ in the age group of 70–80 years. The comparison of results reveals that the epidermis layer

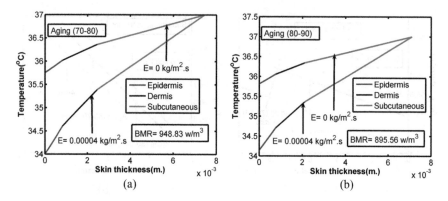

Figure 1.7 Observation of epidermis, dermis, and subcutaneous layers temperatures at $E = 0\,\text{kg/m}^2\text{s}$ and $E = 0.00004\,\text{kg/m}^2\text{s}$ in the elderly with age (a) (60–70) years and (b) (80–90) years.

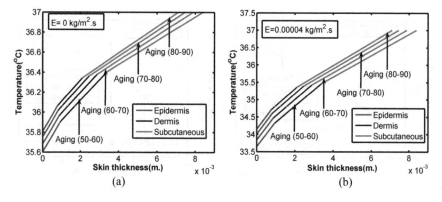

Figure 1.8 Comparison of epidermis, dermis, and subcutaneous tissue temperatures in the age groups of (50–60), (60–70), (70–80), and (80–90) years at (a) $E = 0\,\text{kg/m}^2\text{s}$ and (b) $E = 0.00004\,\text{kg/m}^2\text{s}$.

temperature decreased by 1.74°C, the dermis layer temperature decreased by 1.42°C, and the subcutaneous layer temperature decreased by 0.97°C when the evaporation rate changes to $E = 0.0004\,\text{kg/m}^2\text{s}$ from $E = 0\,\text{kg/m}^2\text{s}$. In the age group of 80–90 years, the temperatures of epidermis, dermis, and subcutaneous tissue reached 35.81°C, 36.06°C, and 36.34°C, respectively, at $E = 0\,\text{kg/m}^2\text{s}$. The temperatures of epidermis, dermis, and subcutaneous tissue decreased by 1.67°C, 1.35°C, and 0.98°C, respectively, at $E = 0.00004\,\text{kg/m}^2\text{s}$ than at $E = 0\,\text{kg/m}^2\text{s}$, as shown in Figure 1.7b. These results delineate that the temperature of each layer of skin decreases with increasing sweat rate due to the loss of heat energy in the form of fluid.

Figure 1.8a and b illustrates the comparison of epidermis, dermis, and subcutaneous tissue temperatures in different age groups. The results indicate that with the age increasing above 50, the epidermis and dermis

Table 1.7 Estimation of epidermis, dermis, and subcutaneous tissue temperatures at different sweat evaporation rates with aging

Age (years)	Steady-state temperature at $E = 0$ kg/m²s			Steady-state temperature at $E = 0.00004$ kg/m²s		
	Epidermis	Dermis	Subcutaneous	Epidermis	Dermis	Subcutaneous
Adults	35.52	35.84	36.44	33.43	34.15	35.52
50–60	35.61	35.90	36.40	33.64	34.32	35.46
60–70	35.69	35.97	36.38	33.84	34.47	35.42
70–80	35.74	36.02	36.36	34.00	34.60	35.39
80–90	35.81	36.06	36.34	34.14	34.71	35.36

layers' temperature increases, while the subcutaneous tissue temperature decreases in both cases $E = 0$ kg/m²s and $E = 0.00004$ kg/m²s. This is due to the reduced thickness of the epidermis and dermis layers and the changing behavior of the subcutaneous tissue with aging.

In the clinical treatment, the body core temperature of people aged 65–74 years was higher than that of people aged 75–84 years. The researchers also suggested that above 85 years, the body core temperature is minimum and ranges from 35.1°C to 36.5°C [18]. The subcutaneous tissue temperature only slightly deviated from these data in our results because the subcutaneous tissue temperature is closer to the body core temperature.

1.7 CONCLUSIONS

The loss of muscle mass reduces the strength and decreases the BMR to play the role of reducing the epidermis and dermis skin thicknesses. The subcutaneous fat tissue increases in sarcopenia, which influences the dissipation of the body temperature. The temperature of the epidermis and dermis layers increases due to decreased thickness. The temperature of subcutaneous tissue decreases due to reduced BMR. These results illustrate that the thicknesses of the skin layers play an important role in the thermoregulation of body. The decrease in BMR due to reduced energy expenditure shows the low significant effects on the temperature of the dermal parts. The sweat evaporation rate also plays a crucial role in controlling the body temperature with aging. Physical therapy and occupational therapy minimize the level of sarcopenia. Regularly performed resistance exercise, resistance training, and physical exercises also maintain muscle mass and muscle strength and increase the BMR which brings on remarkable improvement and changes in aging.

Previously, the researchers experimentally studied the BMR affected by the loss of muscle mass and strength in sarcopenia and suggested treatment options. They haven't studied the mathematical model for the distribution of temperature in the human body due to the change in BMR and thicknesses of skin layers due to sarcopenia. So, this model is prepared to provide a realistic temperature distribution in dermal parts of the body with aging due to the change in thicknesses of skin layers and the effect of basal metabolic rate. This paper aims to maintain physiological parameters in older people who suffer from sarcopenia. It also enables us to develop models that accurately reflect the temperature distribution of infants and pregnant women.

REFERENCES

1. Bianchetti, A. and Andrea, N. 2019. Sarcopenia in the elderly, from clinical aspects to therapeutic options. *Geriatric Care*, 5, 8033–8067.
2. Evans, W. 1997. Functional and metabolic consequences of Sarcopenia. *Symposium: Sarcopenia: Diagnosis and Mechanism*, American Society of Nutritional Science, Pennsylvania State University, University Park, PA.
3. Durnin, J.V.G.A. 1992. Energy metabolism in the elderly. *Nutrition of the Elderly, Nutrition Workshop Series*, 29, 51–63.
4. Kenney, W. and Munce, T. 2003. Aging and human temperature regulation. *Journal of Applied Physiology*, 95, 25982603.
5. Bemben, M.G., Massey, B.H., Bemben, D.A., Boileau, R.A. and Misner, J.E. 1998. Age- related variability in body composition methods for assessment of percent fat and fat-free mass in men aged 20–74 years. *Age and Ageing*, 27, 147–153.
6. Kleiber, M. 1932. Body size and metabolism, Hilgardia. *A Journal of Agricultural Science, California Agricultural Experiment Station*, 6(11), 315–351.
7. Simsek, T.T., Yumin, E.T. and Setal, M. 2019. Aging, body weight, and their effects on body satisfaction and quality of life. *Iranian Red Crescent Medical Journal*, Online. Doi: 10.5812/ircmj.13045.
8. Henry, C.J.K. 2000. Mechanisms of changes in basal metabolism during aging. *European Journal of Clinical Nutrition*, 54(3), S77–S91.
9. Gurung, D.B. 2012. Two dimensional temperature distribution model in human dermal region exposed to low ambient temperatures with air. *Journal of Science, Engineering and Technology, Kathmandu University*, 8(2), 11–24.
10. Acharya, S., Gurung, D.B. and Saxena, V.P. 2013. Effect of metabolic reaction on thermo-regulation in human male and female body. *Applied Mathematics*, 4, 39–48.
11. Pennes, H.H. 1948. Analysis of tissue and arterial blood temperature in resting human forearm. *Journal of Applied Physiology*, 1(2), 93–122.
12. Waalen, J. and Buxbaum, J.N. 2011. Is older colder or colder older? The association of age with body temperature in 18,630 individuals. *Journals of Gerontology Series A: Biomedical Sciences and Medical Sciences*, 66(5), 487–492.

13. Coccarelli, A., Hasan, H.M., Carson, J., Parthimos, D. and Nithiarasu, P. 2018. Influence of aging on human body blood flow and heat transfer. *International Journal for Numerical Methods in Biomedical Engineering*, 34(10), 1–21.
14. Keil, G., Cummings, E. and de Magalhaes, J.P. 2015. How body temperature influences aging and longevity. *Biogerontology*, 16(4), 383–397. Doi: 10.1007/ s10522-015-9571-72.
15. Shrestha, D.C., Acharya S. and Gurung D.B. 2020. Modeling on metabolic rate and thermoregulation in three layered human skin during carpentering, swimming and marathon. *Applied Mathematics*, 11, 753–770.
16. Farage, M.A., Miller, K.W. and Maibach, H.I. 2009. *Text Book of Aging Skin*. Berlin, Heidelberg: Springer- Verlag.
17. Gurung D.B., Saxena V.P. and Adhikari, P.R. 2009. FEM approach to one dimensional unsteady state temperature distribution in human dermal parts with quadratic shape function. *Journal of Applied Mathematics & Informatics*, 27(1–2), 301–331.
18. Gunes, U. and Zaybak, A. 2008. Does the body temperature changes in older? *Journal of Clinical Nursing*, 17(17), 2284–2287. Doi: 10.1111/ j.1365-2702.2007.02272.x.

Chapter 2

Multi-objective university course scheduling for uncertainly generated courses

Sunil B. Bhoi and Jayesh M. Dhodiya
Sardar Vallabhbhai National Institute of Technology

CONTENTS

2.1 INTRODUCTION

Scheduling problem is one of the vital problems in the field of operation research. In particular, schools, colleges, industries, organizations, etc. are facing scheduling problem for assigning activities. For example, scheduling or assigning work means matching people, places, time slots, and facilities. Further, it is not easy to solve problems with plenty of constraints. Usually, constraints are of two types: hard and soft. The problem of faculty course assignment means satisfying all the constraints; for example, every course must be assigned, no two courses to one teacher at one particular time slot, instructor preference to course, courses which are to be assigned between lower and upper limits, all course are assigned as per preferences of teachers as well as the administrator, and preferences based on other criteria such as student convenience and feedback.

In this chapter, we have presented a mathematical model of multi-objective university scheduling for generated courses, when some faculties have left during semester, which has been solved by the fuzzy programming technique using LINGO software.

DOI: 10.1201/9781003303053-3

2.2 LITERATURE REVIEW

University course scheduling is a complex and NP-complete class problem. Many researchers are attracted to carry out research in this field of university scheduling. Bloomfield and McShary [1] applied a heuristic approach in two phases to allocate the courses to teaching faculties based on preferences of teaching faculties. The timetabling or scheduling problem has been solved by different heuristic approaches such as genetic algorithms and tabu search, and expert systems were examined by Costa [2] and Hertz [3]. Badri [4] reported a two-phase approach to allocate courses to teaching faculties and course pair and course pair time slots based on preferences of teaching faculties in two ways. First, the two phases were executed independently, and second, they were executed in combined phase. Kara and Ozdemir [5] also assigned courses to teaching faculties based on preferences by applying a minimax approach. Asratian and Werra [6] extended the basic class teacher model of scheduling and viewed it as a theoretical model. This model captured the most common situations of training programs of universities and schools. It has been seen that this scheduling problem is NP-complete when founded in some sufficient conditions for the existence of a timetable. This work was carried forward by the generalization of the faculty–course assignment problem considered earlier by Ozdemir and Gasimov [7]. They presented and constructed a multi-objective 0–1 nonlinear model for the problem utilizing participants' average preferences and demonstrated an effective way for its solution. To find the optimal schedule, Asmuni et al. [8] presented a fuzzy model. This model was based on the heuristic sorting method. Rachmawati and Srinivasan [9] presented a resource allocation problem as a multi-objective model, and this was a student's project assignment problem. This method used a fuzzy inductive system to develop the model and address the purposes. This model generated optimal allocations, which must satisfy the soft constraints in sequence at some point. To consider some priorities to decide on an agreement among objectives by which the direction of the search path toward attractive regions within the objective space is executed. However, Shatnawi et al. [10] developed a model based on selective courses who opted for the next semester and classified the students. They used a novel clustering technique based on FP-tree. However, Shahvali et al. [11] applied a genetic algorithm to generate schedules. This algorithm in association with local search solved the university course timetable problem. Also, it used the inductive search to solve combined problem and local search, which has the capability of improving efficiency within this genetic algorithm. Chaudhuri and De [12] presented fuzzy genetic heuristic algorithm for university course timetable problem.

It has been observed from the literature that various algorithms are applied to generate scheduling of courses, faculties, and time slots, but there is no model that considers the cases of faculties being transferred, going for some long study leave, or leaving the institute in between semester. Hence, therefore, we consider the case where some faculties have left the institute to study.

2.3 FORMULATION OF PROBLEM

Parameters

Let $M = \{1,2,3,...,m\}$ be the set of courses; $K = \{1,2,...,p\}$ be the set of courses engaged by teaching faculties who have left the institute; $N = \{1,2,3,...,n\}$ be the set of all teaching faculties; and $L = \{1,2,...,l\}$ be the set of existing teaching faculties.

s_i = the number of each of the course i to be offered in a semester.

t_{ij} = the teaching faculty j preference for course i.

u_{ij} = the preference for the course i to teaching faculty j by the administrator.

v_{ij}: = the preference derived based on the result analysis for the allotment of the course i to the teaching faculty j.

w_{ij} = the preference derived based on feedback allotment of the course i to the teaching faculty j.

Decision Variable

Let equation (2.1) define decision variable x_{ij} to represent the assignment of a course to teaching faculty as follows:

$$x_{ij} = \begin{cases} 1, & \text{if } i\text{th course is allotted to } j\text{th teaching faculty,} \\ 0, & \text{otherwise.} \end{cases} \tag{2.1}$$

Constraints

$$\sum_{j=1}^{n} x_{ij} = 1, i = 1,2,3,...,m. \tag{2.2}$$

$$\sum_{i=1}^{m} x_{ij} = 1, \text{for all } j \in L, \text{i.e., existing teaching faculties.} \tag{2.3}$$

$$\sum_{j=1}^{n} x_{ij} = 0, \text{ for all } i \in K, \text{i.e., courses which were allocated} \tag{2.4}$$

to teaching faculties who left the institute.

$$l'_j \le \sum_{i=1}^{m} x_{ij} \le u' \ j = t_j, \text{for each } j = 1,2,3,...,n. \tag{2.5}$$

Equations (2.2)–(2.4) assure that each generated course must be allocated to teaching faculties without disturbing initial courses allocated to teaching faculties. Equation (2.5) does the allocation of each faculty according to their load given. In other words, courses are assigned by calculating their lower as well as upper bounds of their load limit. It will now increase as per load generated as per how many faculties left. It is also assuring that load is distributed or assigned not only by preferences, but also by their load capacity.

Objectives

The mathematical model objectives of the multi-objective university course scheduling problem for generated courses are formulated as follows:

$$L_k(x) = \frac{\sum\limits_{j=1}^{n} x_{ij} t_{ij}}{\sum\limits_{j=1}^{n} x_{ij}}, k = 1,2,\ldots,l \text{ and } i = 1,2,\ldots,m. \tag{2.6}$$

$$A_1 = \frac{\sum\limits_{j=1}^{n}\sum\limits_{i=1}^{m} x_{ij} t_{ij}}{\sum\limits_{j=1}^{n}\sum\limits_{i=1}^{m} x_{ij}}, k = 1,2,\ldots,l \text{ and } i = 1,2,\ldots,m. \tag{2.7}$$

$$A_2(x) = \sum_{i=1}^{m}\sum_{j=1}^{n} u_{ij} x_{ij}. \tag{2.8}$$

$$A_3(x) = \sum_{j=1}^{n}\left(u_j - \sum_{i=1}^{m} x_{ij}\right). \tag{2.9}$$

$$A_4(x) = \sum_{i=1}^{m}\sum_{j=1}^{n} v_{ij} x_{ij}. \tag{2.10}$$

$$A_5(x) = \sum_{i=1}^{m}\sum_{j=1}^{n} w_{ij} x_{ij}. \tag{2.11}$$

Equation (2.5) represents the minimization of preferences of teaching faculties per hour taught for the accepted courses. Equation (2.6) represents the minimization of mean of all preferences of teaching faculties. Equation (2.7) represents the minimization of the preferences of the administrator. Equation (2.8) minimizes the total dispersion from the upper limits of teaching load. Equations (2.9) and (2.10) denote the minimization of preferences based on result and feedback, respectively.

Multi-objective university scheduling problem for uncertainly generated courses (MOUSPUGC)

By using the discussed objective and constraints, the mathematical model is as follows:

Minimize $Z_k = \left[z_1(x), z_2(x), \ldots, z_{n-2}(x), z_{n-1}(x), z_n(x) \right]$,

$$f_k(x) = \frac{\sum\limits_{j=1}^{n} x_{ij} t_{ij}}{\sum\limits_{j=1}^{n} x_{ij}}, \quad k = 1,2,\ldots,(n-5) \text{ and } i = 1,2,\ldots,m, \tag{2.12}$$

$$f_{n-4}(x) = \frac{\sum\limits_{j=1}^{n} \sum\limits_{i=1}^{m} x_{ij} t_{ij}}{\sum\limits_{j=1}^{n} \sum\limits_{i=1}^{m} x_{ij}}, \tag{2.13}$$

$$f_{n-3}(x) = \sum\limits_{i=1}^{m} \sum\limits_{j=1}^{n} u_{ij} x_{ij}, \tag{2.14}$$

$$f_{n-2}(x) = \sum\limits_{j=1}^{n} \left(u_j - \sum\limits_{i=1}^{m} x_{ij} \right), \tag{2.15}$$

$$f_{n-1}(x) = \sum\limits_{i=1}^{m} \sum\limits_{j=1}^{n} v_{ij} x_{ij}, \tag{2.16}$$

$$f_n(x) = \sum\limits_{i=1}^{m} \sum\limits_{j=1}^{n} w_{ij} x_{ij}. \tag{2.17}$$

Constraints

$$\sum_{j=1}^{n} x_{ij} = 1, i = 1,2,3,\ldots,m. \tag{2.18}$$

$$l'_j \le \sum_{i=1}^{m} x_{ij} \le u' \; j = t_j, \text{for each } j = 1,2,3,\ldots,n. \tag{2.19}$$

$$\sum_{i=1}^{m}\sum_{j=1}^{n} x_{ij} = 1 \text{ for all } j \text{ belongs to } L, \text{ i.e., existing teaching faculties.} \tag{2.20}$$

$$\sum_{i=1}^{m}\sum_{j=1}^{n} x_{ij} = 0, \text{ for all } i \text{ belongs to } K, \text{ i.e., courses} \tag{2.21}$$

engaged by faculties who left the institute.

$$x_{ij} = \begin{cases} 1, & \text{if } i\text{th course is assigned to } j\text{th teaching faculty,} \\ 0, & \text{otherwise.} \end{cases} \tag{2.22}$$

2.4 METHODOLOGY

To solve the multi-objective university course scheduling for certainly generated courses, when some teaching faculties left the institute, the fuzzy programming technique is used. Fit a membership function $\mu(Z_k)$ for objective function k for positive and negative ideal solutions. Linear and nonlinear membership functions are utilized to find non-dominated solutions of this university course scheduling problem when some teaching faculties left the institute, and by using these membership functions, the model is converted into the following models.

2.4.1 MOUSPUGC with linear membership function

Equation (2.22) represents the membership function linearly:

$$\mu(Z_k) = \begin{cases} 1, \text{ if } f_k(x) \le l_k, \\ \dfrac{u_k - f_k(x)}{u_k - l_k}, \text{if } l_k < f_k(x) < u_k, k = 1,2,\dots,n \\ 0, \text{if } f_k \ge u_k. \end{cases} \quad (2.23)$$

Using linear membership function, MOUSPCGC is structured as follows:

Maximize λ,
constraints: $\lambda \le \mu (Z_k); k = 1,2,\dots,n.$
And equations (2.11)–(2.21)
$\lambda > 0$ and $\lambda = \min \mu(Z_k).$

2.4.2 MOUSPUGC with nonlinear membership function

Equation (2.23) defines the membership function exponentially:

$$\mu(Z_K) = \begin{cases} 1, \text{ if } f_k(x) \le l_k, \\ \dfrac{e^{-s\psi_k(x)} - e^{-s}}{1 - e^{-s}}, \text{ if } l_k < f_k(x) < u_k, \\ 0, \text{ if } f_k(x) \ge u_k. \end{cases} \quad (2.24)$$

where $\psi_k(x) \le \dfrac{f_k(x) - l_k}{u_k - l_k}, k = 1,2,\dots n$

And s is the non-zero shape parameter. Using exponential membership function, MOUSPCGC is structured as follows:

Maximize λ,
constraints: $\lambda \le \mu (Z_k); k = 1,2,\dots,n.$
And equations (2.11)–(2.21)
$\lambda > 0$ and $\lambda = \min \mu(Z_k).$

2.5 NUMERICAL EXAMPLE

University scheduling is to be done before the commencement of every semester of a program. Due to unexpected circumstances such as teaching faculties transfer, deputation for higher education, or some the teaching faculties leaving the university, scheduling activity is to be re-performed.

Here, one hypothetical data-based example is solved. Let there be six teaching faculties, including one professor, two associate professors, and three assistant professors, and a total of 18 courses, which constitute eight theory courses, two tutorial courses, and eight laboratory courses.

Let $M = \{01,02,03,04,05,06\}$;

$N = \{1,2,\ldots, 18\}$;

$K = \{01,04,11,06,13,18\}$;

$L = \{02,06\}$

Table 2.1 represents the teaching load boundaries for teaching faculties (Tables 2.2–2.5).

2.6   RESULTS AND DISCUSSION

Utilizing the data provided in table 2.1 to 2.5 of the previous section, a model has been formulated and solved by the fuzzy programming technique. The solutions are found using LINGO software. Table 2.6 represents

Table 2.1 Upper and lower limits of teaching faculties load

Teaching faculties	I	II	III	IV	V	VI
Upper limit	3	4	4	5	5	5
Lower limit	2	3	3	3	3	3

Table 2.2 Teaching faculties preferences for courses

Teaching faculties	P1	P2	P3	P4
I	03,10,12,14,17	02,06	04,11	08,15
II	09,18	01,04,05,11,15	07,13,16	-
III	08,13	14,16	02,12	03,06,17,18
IV	15	11,18	05,	01,04,07,09,13
V	07,18	08,12	04,10,16	02,05,14
VI	06,13	01,03,09	18	17

Table 2.3 Administrator preferences for teaching faculties and courses

Teaching faculties	P1	P2	P3	P4
I	02,06,10	03,04,11	08,15,17	12,14
II	01,15,16	05,13	04,09	07,11,18
III	06,14,16	08,10	02,12,17	03,13,18
IV	04,15	05,13	07,11,17	01,09
V	07,08	02,10,18	04,12	05,14,16
VI	01	03,18	09,13	06,17

Table 2.4 Preferences for courses to teaching faculties based on feedback

Teaching faculties	P1	P2	P3	P4	P5
I	06,08	02,10	12,14	03,11	04,15,17
II	04,11	01,13	05,15	07,16	09,18
III	08,12	06,13	16,17	02,14	03,10,18
IV	01	04,17	05,15	07,13	09,11
V	02,10	04,12	05,14	07,16	08,18
VI	09,17	13,18	06	03	01

Table 2.5 Preferences for courses to teaching faculties based on result

Teaching faculties	P1	P2	P3	P4	P5
I	03,10	06,11	08,14	02,12	04,15,17
II	04,18	01,09,11	13	05,15	07,16
III	06,16	10,18	02,12	08,14	03,13
IV	01,09,11	07,13	05,15	04	17
V	05,18	08,14	07,10	02,12	04,16
VI	17	18	06	03,13	01,09

Table 2.6 Objective values for objective functions

Objective functions	F1	F2	F3	F4	F5	F6	F7	F8	F9	F10	F11
$\lambda = 0.5652173$	0.8	2	1.421053	2.125	1.9325	1.8	2.1875	27	1	50	46

Table 2.7 Courses allocated using the linear membership function

Teaching faculties	Courses allotted using LMF
I	02,12
II	01,04,11
III	03,08,16
IV	09,15,17
V	05,07,10,14
VI	06,13,18

the objective values for objective functions and table 2.7 represents initial courses allotted (Table 2.7).

Suppose that teaching faculties II and IV have left the institute in between the semester. The generated courses are allotted to existing teaching faculties.

Table 2.8 represents the optimal values of the objective functions. Table 2.9 represents the generated courses allotted to existing teaching faculties. The overall level of satisfaction is $\lambda = 0.5357143$. Initially, all courses are allocated to the teaching faculties with an overall level of satisfaction $\lambda = 0.5652173$. After modification, the generated courses are allocated without disturbing initial allocations. Also, it can be solved by fitting the nonlinear membership function and distinct allocations are arrived by varying the shape parameter s. This model provides a quick and effective solution. Figures 2.1 and 2.2 represent the objective values.

Table 2.8 Objective values for objective functions

Objective functions	F1	F3	F4	F5	F6	F8	F9	F10	F11
$\lambda = 0.5357143$	1	1.32	2.142857	1.9325	1.6818	25	9	50	51

Table 2.9 Generated courses allotted to existing teaching faculties

Teaching faculties	Courses allotted using LMF
I	02,06,12
III	03,08,13,16
IV	01,04,09,11,15,17
V	05,07,10,14,18

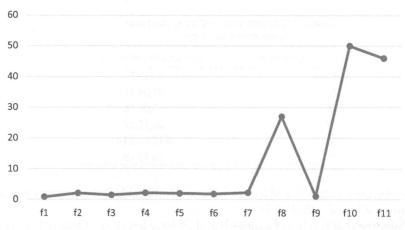

Objective values for objectives

Figure 2.1 Objective values for objectives.

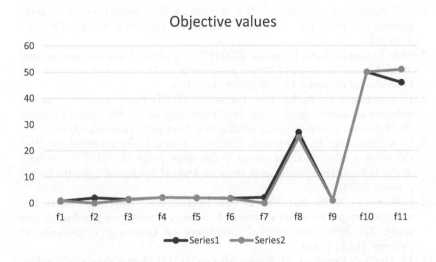

Figure 2.2 Objective values.

2.7 CONCLUSIONS

This chapter presents a mathematical model of multi-objective university scheduling problem for generated courses, when some teaching faculties left the institute. This model is viewed as 0–1 integer programming. To develop this model, preferences from teaching faculties, the administrator, and learners are obtained. This is a modified version of the basic model, including calculated changes in lower and upper limits accordingly generated courses with disturbing initial allocations. To demonstrate and validate the proposed model, one numerical case based on hypothetical data is presented. The efficient solutions are obtained by the fuzzy programming technique, and the results are found using LINGO19.0 software.

REFERENCES

1. S.D. Bloomfield, M.M. McShary, (1979) Preferential course scheduling, *Interfaces* 9(4) 24–31.
2. D. Costa, (1994) A tabu search algorithm for computing an operational timetable, *European Journal of Operational Research* 76, 98–110.
3. A. Hertz, (1991) Tabu search for large scale timetabling problems, *European Journal of Operational Research* 54, 39–47.
4. M.A. Badri, (1996) A two stage multiobjective scheduling model for faculty-course assignments, *European Journal of Operational Research* 94 16–28.
5. I. Kara, M.S. Ozdemir, (1997) Minmax approaches to faculty course assignment problem, *Proceedings of the 2nd International Conference on the Practice and Theory of Automated Timetabling*, Toronto, 167–181.

6. A.S. Asratian, D. Werra, (2002) A generalized class teacher model for some timetabling problems, *European Journal of Operational Research* 143, 531–542.

7. M.S. Ozdemir, R.N. Gasimov, (2004) The analytic hierarchy process and multiobjective 0–1 faculty course assignment problem, *European Journal of Operational Research* 157 (2) (2004) 398–408.

8. H. Asmuni, E.K. Burke, J.M. Garibaldi, (2005) Fuzzy multiple heuristic ordering for course timetabling, *The Proceedings of the 5th United Kingdom Workshop on Computational Intelligence (UKCI05)*, London, 302–309.

9. L. Rachmawati, D. Srinivasan, (2010) A hybrid fuzzy evolutionary algorithm for a multi-objective resource allocation problem, *IEEE Proceedings of the Fifth International Conference on Hybrid Intelligent Systems*, Rio de Janeiro, 2005.

10. S. Shatnawi, K. Al -Rababah, B. Bani-Ismail, (2010) Applying a novel clustering technique based on FP-tree to University timetabling problem: A case study. *The 2010 International Conference on Computer Engineering & Systems*, IEEE, Cairo.

11. M. Shahvali Kohshori, M. Saniee Abadeh, (2012) Hybrid genetic algorithms for university course timetabling, *IJCSI International Journal of Computer Science* 9(2), 446–455.

12. A. Chaudhuri, D. Kajal, Fuzzy genetic heuristic for university course timetable problem, *International Journal of Advances in Soft Computing and its Applications* 2(1), ISSN 2074-8523.

Chapter 3

MChCNN: a deep learning approach to detect text-based hate speech

Pallavi Dash, Namrata Devata, and Debani Prasad Mishra
International Institute of Information Technology Bhubaneswar

CONTENTS

3.1 INTRODUCTION: BACKGROUND AND DRIVING FORCES

All around the globe, we tend to see a worrying irruption of social phobia, racism and intolerance together with rising racism, anti-Muslim emotion and injustice to Christians. Social networking being exploited as platforms for zealotry, Neo-Nazi and racism movements are on the march. The public talk is being weaponized for political gain in a combustible manner with discussions that describe and automatize minorities, travelers, exiles and ladies. Addressing hate speech does not imply restricting or banning free speech. It means that hate speech should be discouraged from thickening and inflicting damage to society, notably provocation to discrimination, hostility and violence, which is prohibited under law [1,2].

The darkness and adaptability of the network have made it possible for people to talk in an offensive way. Due to the huge scale of the Internet, the requirement for ascendable, automatic strategies of hate speech detections has fully grown well. So, "what is precisely hate speech?" [3].

We could very well summarize from paper [4] that hate speech is an expression that is probably going to cause anguish or antagonize alternative people on the premise of their association with a specific cluster or incite hostility toward them. A keynote to think about when defining hate speech is that it should not be paired with repulsive terms. Since anyone who uses offensive language is not mechanically violent, for all sorts of purposes, people prefer to use words that are highly offensive, while some with roguish means, some in qualitatively different ways. For instance, in common language, individuals use certain words, whether they quote rap lyrics once, or maybe for fun. On social media, such a language is present, making it essential to any detector of usable hate speech. Hate speech could have an effect directly or indirectly on individuals' psychological well-being, with the number of harms considerably larger just in case of exploitation, compared to mere witnessing. Victims of online hate speech could develop social restlessness and feel the requirement to detach themselves from their encompassing.

There have been an increasing number of hate speech identification inquiries in recent years. The uprising of the pandemic COVID-19 has led to the increase in social media use, and folks from all age clusters have become terribly active on the web, which has led to the rise in text generation. We could deduce from paper [5] that since the end of 1940s, the study on natural language processing (NLP) has been in progress. One of the main computer-based programs dealing with natural languages was machine translation (MT). The information science research advanced from the age of punch cards and execution of instruction, during which it could take up to seven minutes to interpret sentences, to the age of Google, and therefore the likes of it during which several websites are processed for a second time. The issue was specifically identified by current methods as a challenge for supervised text classification. We may infer from paper [6] that these may be divided into two categories: One relies on manual feature engineering, which is then used by algorithms such as SVM, naive Bayes and regression supply (classic techniques), and the other may be a newer approach of deep learning that uses neural networks to mechanically learn abstract multi-layer features from data (deep learning techniques).

In this chapter, we have proposed a multichannel convolutional neural network where the extracted features are used to differentiate hate speech from non-hate speech. Vectors of words relying on linguistics data are designed for all the tokens from the dataset using an algorithm for unsupervised learning, GloVe. The word vectors are merged into a collection of extracted attributes, downsized by max pooling, and three different n-grams are concatenated and fed into the neural network model in conjunction with the character to figure out the groups of each tweet. This chapter shows the comparison between our planned model and the previous implementations with a benchmark dataset.

3.2 RELATED WORK

Nemanja Djuric et al. [7] proposed a two-step process for the identification of speech concerned to hate content. The authors implemented paragraph2vec in their architecture, which focuses on the combined modeling of various comments and words, where it is observed that the representations are distributed in a common space using a continuous BOW (CBOW) neural language architecture. As a result, a low-dimensional text embedding is established, where numerous sentences and phrases are found to have semantic relationships. The embeddings are then integrated to train a binary classifier to determine the difference between the clean and hateful remarks. The accuracy delivered by the proposed model is 80.7%.

Edel Greevy et al. [8] presented a description of the methods used to build and test a text categorization scheme for the automated categorization of texts concerned with racism. They addressed the issue of detecting racial texts using a trained support vector machine (SVM). The smallest test set performed better for the representation of POS achieving an accuracy of 91.67% by using the polynomial kernel function. There was no formal comparison of the study of character-based functionality for the identification of abusive language even though the character n-grams and word had been used as separate features in various applications of natural language processing. Yashar Mehdad et al.'s [9] research work included the study of the usefulness of features for offensive language identification in the comments available on the Internet, generated hugely by users. The focus of their work demonstrated that such methodologies outperformed previous state-of-the-art and other clear baseline approaches. They explored the role that character n-grams perform in this problem by investigating the implementation in two separate algorithms. On comparing the outcomes of both the state-of-the-art methods by examining on a corpus, they summarized the following: (i) character n-grams outperform and deliver better results in comparison with word n-grams in both algorithms, and (ii) the architectures suggested in their research outperform the previously implemented state-of-the-art methods on their dataset. They adopted multiple supervised classification approaches with lexical and morphological characteristics to assess the different features of the user's comments. The biggest difference between their classification process and their previous work in this field is that they used a hybrid approach based on discriminatory and generative classifiers. They also examined three separate approaches for identifying militant languages. The first is based on the distributional representation of comments (C2V), which is supposed to be a good starting point for the project. The next two, RNNLM and NBSVM, were used as methodologies for determining the effect of character-based vs. token-based functionality. He concluded that C2V and n-gram tokens were not successful. The overall scorer baseline and current state-of-the-art were defeated by NBSVM using character n-gram (77 F1 and 92 AUC). This indicated that it's a lightweight approach that uses character-level features and will outperform approaches that are more intricate.

Badjatiya et al. [10] performed broad tests with various neural network designs to learn logical representations and semantic word embeddings in order to deal with the hate speech recognition complexity. They investigated a benchmark dataset of 16,000 explained tweets, and the deep learning techniques utilized beat conventional word n-gram methods by around 18 in the F1 scores.

Gambäck et al. [11] designed a classifier that assigns each tweet to one of four limited classifications: prejudice, sexism, both (bigotry and sexism), and non-hate speech. They trained four convolutional neural network architectures on respective 4-gram character word vectors dependent on semantic data employed using the word embedding word2vec that arbitrarily produced word vectors, and these word vectors joined with character n-grams. The max pooling layer was used to efficiently downsize the feature set in the networks, and for the characterization of the tweets, softmax was used in the output.

Zimmerman et al. [12] demonstrated how significantly various weight initialization strategies contributed while considering any research using deep learning techniques. This paper highlighted the basic ensemble strategy for neural networks, which measurably improved over a single model. Moreover, the authors indicated that individual architectures have a huge difference when contrasted with the variance of ensemble models. It is also observed that, in comparison with the mean of the sub-model, the ensemble model provides better results on the test sets.

3.3 EXPERIMENTS AND RESULTS

For the execution of our multichannel neural network architecture model, we have used a hate speech dataset from Kaggle that contains 31,962 tweets from the microblogging website Twitter. The dataset contains around 27,456 non-hate speech vs 5,234 hate speech. The annotation given by the author of the dataset for hate speech is 1, whereas for non-hate speech, it is 0.

The dataset collected contains a huge set of text-based streams of tweets that needs pre-processing before being fed into the neural network. In our experiment, to achieve best results, we have cleaned the tweets before tokenization of the words. Using the NLTK library, we have removed the punctuations, converted all the words to lowercase, eliminated all the URLs and removed the symbols before usernames and hashtags. The texts have also undergone lemmatization for effective morphological analysis.

3.3.1 Input layer

Initially, the convolution neural network was designed to process various image-based data that possess constant size and are of low dimensionality. Therefore, the network was fed with inputs of width multiplied with height multiplied with 3. Lately, in papers [13,14], the authors have explored the implementation of CNNs with modifications that can take sentences converted into arrangement of one-hot vectors.

3.3.2 Embedding layer

While processing text data, before feeding them into the neural network, it is essential to convert them into vectors. To retain the relationship between the vectors formed, the embedding layer helps to hold words with similar meaning in the relatively close context. In papers [15,16], several researchers have explored the efficiency of word embeddings mechanism to enhance text classification.

After tokenization, each of the words is converted to tokens, which are remodeled into word embeddings, and hence as a result, words holding similarity in semantics are mapped next to each other. These efficiently apprehend the likeness of the terms. For our implementation, we have integrated the predictive method of identifying the similarity between words. Global Vectors (GloVe) for word representation [17,18] is the subsequent technique that attempts to foresee a word from its neighbors regarding learned small, dense embedding vectors.

3.3.3 Convolutional layers

In most of the deep learning architectures, for efficient classification of text, a word embedding layer is combined with a single-dimensional convolutional neural network. From Figure 3.1, we can see that in our experiment, we have ameliorated the architecture by integrating three parallel convolutional neural networks that take the tweets using kernels of multiple magnitudes such as 4, 5 and 6. The multichannel convolutional neural network [19,20] created reads text streams with different n-gram sizes. The general architecture used for categorization includes the implementation of an embedding layer as an input to the neural network, which is accompanied by a layer of single-dimensional convolutional neural network. Convolutional layer is then followed by a max pooling layer and concluded with a prediction output layer. In the output layer, we have used the softmax activation function [21,22] for the binary classification, which is given by equation (3.1):

$$P\left(y = j \mid \Theta^{(i)}\right) = e^{\Theta^{(i)}} \Big/ \sum_{j=0}^{k} e^{\Theta_k^{(i)}} \tag{3.1}$$

where

$$\Theta = w_0 x_0 + w_1 x_1 + \ldots + w_k x_k = \sum_{i=0}^{k} w_i x_i = w^T x \tag{3.2}$$

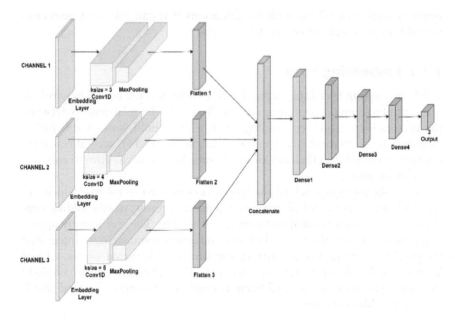

Figure 3.1 Model architecture.

The size of the kernel in the convolutional layer defines the number of words to be considered as the convolution is passed across the input text stream, providing a grouping parameter. In our model, we have used multiple versions of the general architecture with different sized kernels. Before feeding into the dense layer, all the three versions are concatenated. This allows the text-based streams to be analyzed at various intentions or different groups of words at a time. It enables the model to extract the best integrated representations.

To achieve the optimized results from our model, it is necessary to split our testing dataset into testing and validation datasets. Until the completion of the model, it is crucial to see how the architecture works by using a validation dataset.

Figure 3.2 shows the accuracy achieved by the training and testing datasets. From the graph, we can observe whether the model is overfitting or under-fitting. Looking at both the plots, we can see that our model has learnt the distinguishing features well and has performed well by delivering around 95% on the unseen testing dataset with a minimum loss. The batch size used for our experiment is 16 with a total of 50 epochs. We have used the binary cross-entropy loss for our classification algorithm coupled with an Adam optimizer and a learning ratio of 10^{-4}.

From Figure 3.3, we can infer the loss incurred during the validation and testing of the model. The model shows a slight increase in the loss during the testing phase of the model. It is highly recommended to collect more

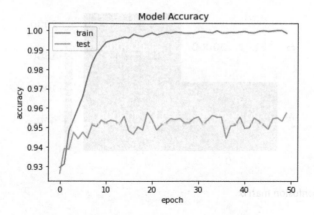

Figure 3.2 Model accuracy vs epoch.

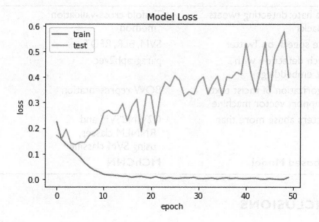

Figure 3.3 Model loss vs epoch.

data to reduce overfitting. For carrying out our experiment, it should be noted that we have chosen a smaller set of data from the Kaggle website.

To analyze our output, we have also generated a confusion matrix [23,24], which is shown in Figure 3.4, and recorded evaluation scores, such as F1, recall and precision [25]. The performance of the model on the test data is validated using the confusion matrix. In Figure 3.4, we can also observe that a total of 101 tweets were wrongly classified by the model out of 2,397 unseen tweets. Our model scored an F1 score of 0.67, recall of 0.58 and precision score of 0.78. Table 3.1 gives an overall insight into the accuracies achieved by various state-of-the-art models along with the performance of our proposed architecture.

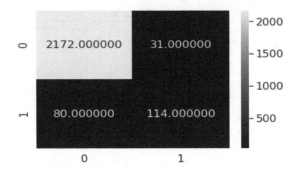

Figure 3.4 Confusion matrix.

Table 3.1 Comparison of SVM-RBF, Poly and MK classification accuracy on the applied dataset

Research title	Method	Accuracy
Locate the hate: detecting tweets against blacks	Tenfold cross-validation method	0.76
Cyber hate speech on Twitter	SVM, BLR, RFDT	0.77
Hate speech detection with comment embeddings	paragraph2vec	0.8
Text categorization of racist texts using a support vector machine	BOW representation	0.86
Do characters abuse more than words?	C2V, NBSVM and RNNLM classes, using SVM classifier	0.93
Our Proposed Model	**MChCNN**	**0.95**

3.4 CONCLUSIONS

In this chapter, our objective is to analyze hate speech and propose suitable and efficient models for its effective detection. An important aspect of successful determination of hate speech is being able to understand firsthand what hate speech is, which is included in our research work. From the related research work conducted in the domain, it was observed that deep learning architectures have significantly outperformed machine learning approaches. It was also observed that with the integration of various pre-processing techniques in the natural language, better results have been achieved. A multichannel neural network model for the efficient detection of hate speech is proposed. The model could capture the word relations in the text representations from its multiple n-gram convolutional filters. To the best of our understanding, our implementation with the included hyper-parameter range has yielded finer results than many state-of-the-art models.

In future, researchers must start concentrating on errors in the dataset for improvised output. While investigating the incorrect predictions in the hate category, this could help with clarification why this category is difficult to predict and it will be insightful to explore the terminology useful for distinguishing between offensive and hate speech. More dense networks and data augmentation techniques can be employed and experimented to effectively recognize the distinguishing features that can predict hate speech successfully.

REFERENCES

1. S. Yuan, X. Wu, and Y. Xiang. "A two phase deep learning model for identifying discrimination from tweets" In *Proceedings of 19th International Conference on Extending Database Technology*, pp. 696–697, 2016. Doi:10.5441/002/edbt.2016.92

2. D. P. Mishra and P. Ray, "Fault detection, location and classification of a transmission line," *Neural Computing and Applications*, 2018, Vol. 30, No. 5, pp. 1377–1424. Doi: 10.1007/s00521-017-3295-y.

3. M. Yashar, and J. Tetreault. "Do characters abuse more than words?" In *Proceedings of the 17th Annual Meeting of the Special Interest Group on Discourse and Dialogue*, pp. 299–303, 2016. Doi: 10.18653/v1/W16-3638.

4. G. Björn, and U.K. Sikdar. "Using convolutional neural networks to classify hate-speech." In *Proceedings of the First Workshop on Abusive Language Online*, Co-Funded by the Rights, Equality and Citizenship Programme of the European Union (2014–2020), 2017, pp. 85–90.

5. R. Mishra, and D. P. Mishra. "Comparison of neural network models for weather forecasting," *Advances in Energy Technology Proceedings of ICAET*, Jan. 2020, Vol. 2020, pp. 79–89.

6. G. Xiang, B. Fan, L. Wang, J. Hong, and C. Rose. "Detecting offensive tweets via topical feature discovery over a large scale Twitter corpus." In *Proceedings of the 21st ACM International Conference*. CIKM'12, October 29–November 2, 2012, Maui, HI.

7. N. Djuric, J. Zhou, R. Morris, M. Grbovic, V. Radosavljevic, and N. Bhamidipati. "Hate speech detection with comment embeddings." In *Proceedings of the 24th International Conference on World Wide Web - WWW'15 Companion*, 2015. Doi: 10.1145/2740908.2742760.

8. E. Greevy, and A.F. Smeaton. "Classifying racist texts using a support vector machine." In *Proceedings of the 27th Annual International Conference on Research and Development in Information Retrieval- SIGIR'04*, 2004. Doi: 10.1145/1008992.1009074.

9. Y. Mehdad, and T. Joel. "Do characters abuse more than words?" In *Proceedings of the 17th Annual Meeting of the Special Interest Group on Discourse and Dialogue*, pp. 299–303, 2016. Doi: 10.18653/v1/W16-3638.

10. P. Badjatiya, G. Shashank, G. Manish, and V. Vasudeva. "Deep learning for hate speech detection in tweets." In *Proceedings of the 26th International Conference on World Wide Web Companion*, Perth, 2017, pp. 759–760.

11. B. Gambäck, and U.K. Sikdar. "Using convolutional neural networks to classify hate-speech." In *Proceedings of the First Workshop on Abusive Language* online, Vancouver, BC, Association for Computational Linguistics, 2017, pp. 85–90.

12. S. Zimmerman, U. Kruschwitz, and C. Fox. "Improving hate speech detection with deep learning ensembles." In *Proceedings of the Eleventh International Conference on Language Resources and Evaluation (LREC 2018)*, Miyazaki, European Language Resources Association (ELRA), 2018.

13. P. Ray and D.P. Mishra. "Application of extreme learning machine for underground cable fault location," *International Transactions on Electrical Energy Systems*, December 2015, Vol. 25, No. 12, pp. 3227–3247.

14. P. Badjatiya, G. Shashank, G. Manish, and V. Vasudeva. "Deep learning for hate speech detection in tweets." In *Proceedings of the 26th International Conference on World Wide Web Companion*, 2017, pp. 759–760.

15. P. Wang, B. Xu, J. Xu, G. Tian, C.-L. Liu, and H. Hao. "Semantic expansion using word embedding clustering and convolutional neural network for improving short text classification." *Neurocomputing*, 2016, Vol. 174, pp. 806–814.

16. S.K. Panda, P. Ray, and D.P. Mishra. "A study of machine learning techniques in short term load forecasting using ANN." *Intelligent and Cloud Computing. Smart Innovation, Systems and Technologies*, December 2019, Vol. 194, pp. 49–57, Springer, Singapore. Doi: 10.1007/978-981-15-5971-6_6.

17. Z.H. Kilimci, and S. Akyokuş. "The evaluation of word embedding models and deep learning algorithms for Turkish text classification." In *2019 4th International Conference on Computer Science and Engineering (UBMK)*, IEEE, Piscataway, NJ, 2019, pp. 548–553. http://dx.doi.org/10.1109/UBMK.2019.8907027

18. B. Guo, C. Zhang, J. Liu, and X. Ma. "Improving text classification with weighted word embeddings via a multi-channel TextCNN model." *Neurocomputing*, 2019, Vol. 363, pp. 366–374.

19. S. Panda, D.P. Mishra, and S.N. Dash. "Comparison of ANFIS and ANN techniques in fault classification and location in long transmission lines." In *2018 International Conference on Recent Innovations in Electrical, Electronics & Communication Engineering (ICRIEECE)*, Bhubaneswar, India, 2018, pp. 1112–1117. Doi: 10.1109/ICRIEECE44171.2018.9008605.

20. M. Andre, and R. Astudillo. "From softmax to sparsemax: A sparse model of attention and multi-label classification." In *International Conference on Machine Learning*, 2016, pp. 1614–1623.

21. P. Ray, D. P. Mishra, K. Dey and P. Mishra, "Fault detection and classification of a transmission line using discrete wavelet transform & artificial neural network." In *2017 International Conference on Information Technology (ICIT)*, Bhubaneswar, India, 2017, pp. 178–183. Doi: 10.1109/ICIT.2017.24.

22. X. Liang, X. Wang, Z. Lei, S. Liao, and S.Z. Li. "Soft-margin softmax for deep classification." In *International Conference on Neural Information Processing*, Springer, Cham, 2017, pp. 413–421.

23. S. Visa, B. Ramsay, A.L. Ralescu, and E.V.D. Knaap. Confusion matrix-based feature selection. *MAICS*, 2011, Vol. 710, pp. 120–127.

24. P. Ray, and D.P. Mishra, "Artificial Intelligence based fault location in a distribution system." In *13th International Conference on Information Technology*, Bhubaneswar, India, 22nd–24th December, ICIT 2014, pp. 18–23.
25. P. Flach, and M. Kull. "Precision-recall-gain curves: PR analysis done right." In *Advances in Neural Information Processing Systems*, Massachusetts Institute of Technology (MIT) Press. 2015, pp. 838–846. https://papers.nips.cc/paper/5867- precision-recall-gain-curves-pr-analysis-done-right

Chapter 4

PSO-based PFC Cuk converter-fed BLDC motor drive for automotive applications

M. Ramasubramanian and M. Thirumarimurugan
Coimbatore Institute of Technology

CONTENTS

4.1 INTRODUCTION

A brushless direct current (BLDC) motor is advantageous for low- and medium-power and automotive applications due to better efficiency, better flux density, low requirements, reduced harmonics and a wide range of speed [1]. These advantages are helpful in various applications in household appliances, transportation, aerospace, HVAC lines, robotics, etc. [2]. Normally, a BLDC motor is a 3Ø synchronous motor having 3Ø concentrated stator windings and a permanent magnet in rotor windings [3]. It does not have brushes and also a commutator. The posture of the rotor signals is determined by the Hall effect sensor, which is electronically commutated over the mechanical commutation [4,5]. In a power factor correction (PFC), converter-fed BLDC motor varying the perpetual DC link voltage for controlling the speed through VSI for duty cycle ratio of greater frequency pulse-width-modulated (PWM) signals is presented [6,7]. The buck converter's operation output voltage was lower than the input voltage, and the VSI worked in high switching losses for improving the efficiency [8]. The buck converter was mainly used in low-power applications. In a boost converter, the operation output voltage was more than the input voltage, which was based on direct torque-controlled BLDC motor and VSI acting at high switching losses [9]. The boost converter was used in medium-power applications. The configuration of bridgeless converters such as buck-boost, SEPIC and zeta converters contains a front-end converter that increases the

DOI: 10.1201/9781003303053-5

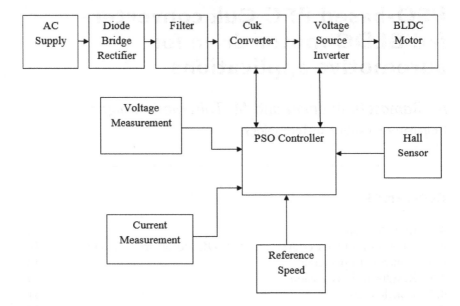

Figure 4.1 System block diagram.

power factor at the AC mains [10]. Here, we are using the Cuk converter to improve the power factor of the BLDC drive (Figure 4.1).

The particle swarm optimization (PSO) was invented by Kennedy and Eberhart to simulate the motion of swarms of birds as portion of a socio-cognitive learning. Each particle is searching for the optimum position. Therefore, each particle is moving toward the optimum position and hence has a velocity. Each particle reminisces the position it was in where it had its most excellent result so far (its personal best). Each particle in the swarm cooperates each other. They swap evidence concerning what they've learnt in the locations they have visited. A particle has a locality associated with it. A particle is well known with the fitness of those in its locality and moves toward the posture of the one with best fitness. This posture is merely utilized to fine-tune the particle's velocity. The above lines are the brief introduction about the PSO controller.

4.2 OPERATION OF CUK CONVERTER-FED BLDC MOTOR DRIVE SYSTEM

Figure 4.2 displays a PFC Cuk converter-fed BLDC motor drive operated in the discontinuous conduction method (DCM) and the continuous conduction method (CCM) using a PSO controller. The Cuk converter is

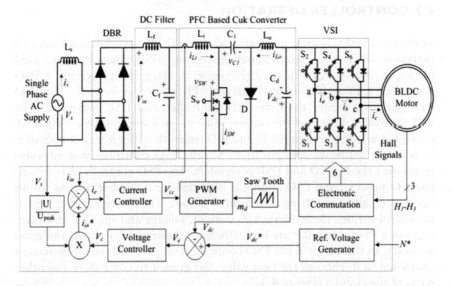

Figure 4.2 Circuit diagram for the proposed system.

advantageous over buck and boost converters due to VSI working in low switching losses in the Cuk converter. In Cuk converter, the output voltage is less than or greater than or equal to the input voltage and also it is used in both low- and medium-power applications.

A MOSFET is employed in a Cuk converter in favor of PFC and controlling the DC link voltages. An insulated-gate bipolar transistor (IGBT) in voltage source inverter (VSI) is utilized for minimal switching losses and also for electronic commutation of the BLDC motor. The operation proceeds with the input single-phase AC supply, and a diode bridge rectifier (DBR) is employed to convert the alternating current to direct current. The DC supply is given to the Cuk converter through an LC filter for reducing the harmonics. The Cuk converter works in the discontinuous conduction method (DCM) and the continuous conduction method (CCM), and it varies the direct current link voltage. The direct current voltage is given to the VSI at 180° mode. The output AC voltage is given to the BLDC motor to operate at overspeed. The Hall sensors are utilized to measure the rotor signals. The Hall sensor signal was compared to the reference signal from the controller. The controller controls the speed of the BLDC motor and PWM signals to the Cuk converter for controlling the current and voltage for enhancing the power factor at the AC mains of the circuit with a broad range of speed control.

4.3 CONTROLLER OPERATION

In conventional schemes, proportional integral (PI), proportional integral derivative (PID) and digital signal processing (DSP) controllers are used for improving system efficiency (Figure 4.3). Here, in our system, a PSO controller with a proportional integral (PI) controller is used. The PSO controller was better than the PID controller and DSP controller. The PID controller decreased the output value at each step response. The DSP controller was better than the PI and PID controllers, but it was not possible in working hardware configuration. So, the PSO controller was advantageous over the PI, PID and DSP controllers. The PSO controller has a comparative process, and PSO coding is obtained for a number of runs for an optimized value for getting the required output. The PSO controller is used to control various elements such as current, voltage and speed. The advantages of a PSO controller are high efficiency, better performance and good accuracy. In this chapter, the PSO controller compares all the measurement values fed to it, chooses the best value and gives it to controllers and other parts of the circuits (Figure 4.4).

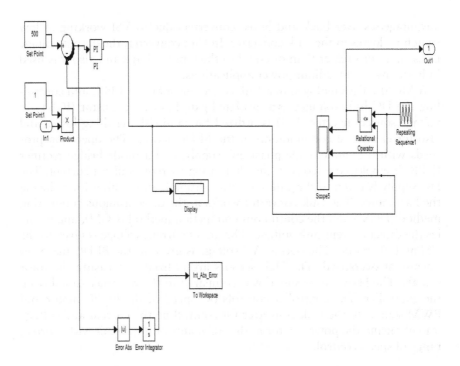

Figure 4.3 PSO controller simulation circuit.

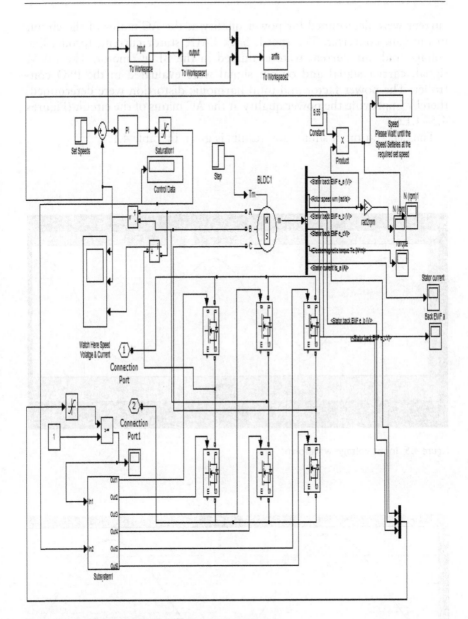

Figure 4.4 BLDC motor with VSI simulation circuit.

4.4 RESULTS AND DISCUSSION

The execution of the Cuk converter-fed BLDC motor drive based on PSO was simulated in Simulink/MATLAB software. The execution performance was determined based on several parameters. The input voltage and input current were determined for power quality at the AC mains of the circuit in the Cuk converter. The speed, back EMF, stator current, torque, line voltage and line current were evaluated in the BLDC motor. The PWM signal, carrier signal and output signal were evaluated in the PSO controller. The power factor and total harmonic distortion were determined, thereby improving the power quality at the AC mains of the circuit (Figures 4.5–4.12).

The declaration of simulation results is given in Table 4.1.

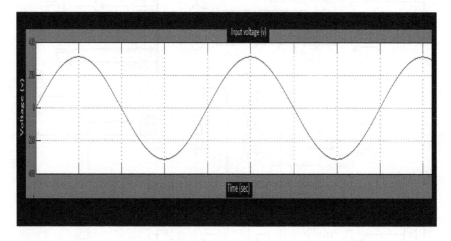

Figure 4.5 Input voltage waveform.

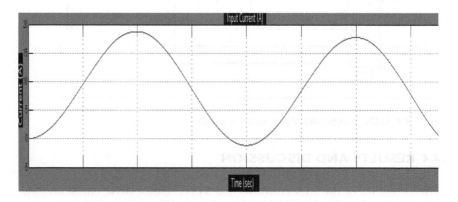

Figure 4.6 Input current waveform.

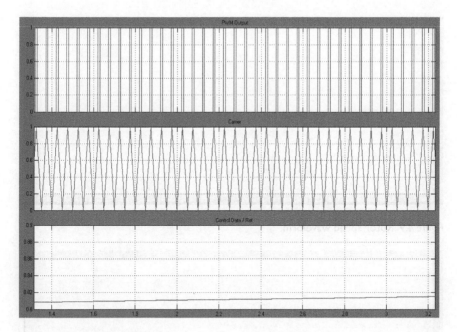

Figure 4.7 PSO controller output.

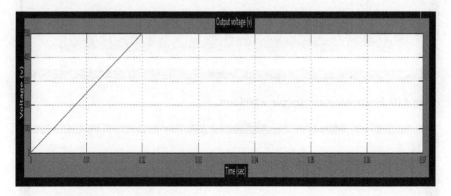

Figure 4.8 Output voltage waveform.

4.5 CONCLUSIONS

A Cuk converter-fed BLDC motor drive based on PSO was designed to improve the power factor at the AC mains for low-power equipment such as pumps, blowers and cooling fans. The speed of the BLDC motor was well ordered (controlled) by changing the DC link voltage of VSI for low switching losses. The power factor value in the proposed system (0.98) was

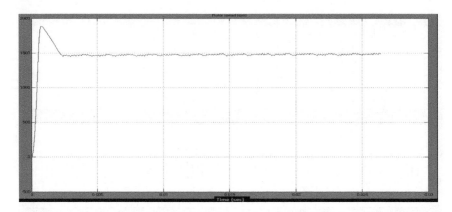

Figure 4.9 Rotor speed waveform.

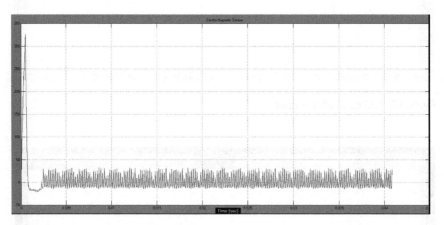

Figure 4.10 Electromagnetic torque waveform.

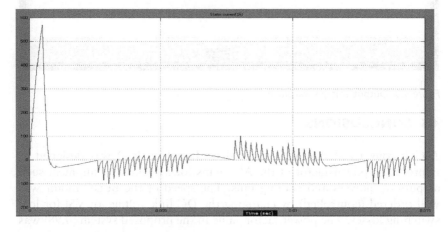

Figure 4.11 Stator current waveform.

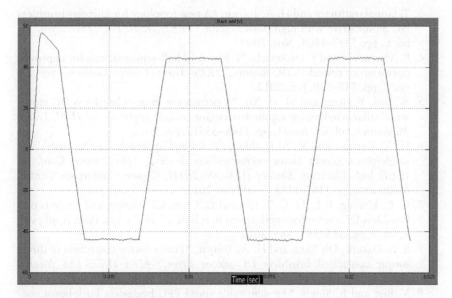

Figure 4.12 Back EMF waveform.

Table 4.1 Declaration of simulation results

S. No.	Parameters	Measurement values
I	Input voltage	230V
2	Input current	18A
3	Output voltage	500V
4	Motor speed	1,500rpm
5	Total harmonic distortion	2.6%
6	Power factor	0.98

better than in the existing system [1]. The output was evaluated by using Simulink/MATLAB software. Finally, the proposed system gives a satisfactory result and also suggests solutions for low-power BLDC motor drives.

REFERENCES

1. V. Bist and B. Singh, "An adjustable speed PFC bridgeless buck-boost converter fed BLDC motor drive", *IEEE Trans. Ind. Electron.*, vol. 61, no. 6, pp. 2665–2677, Jun. 2014.
2. V. Nasirian, Y. Karimi, A. Davoudi and M. Zolghadri, "Dynamic model development and variable switching frequency control for DCVM Cúk converters in PFC applications," *IEEE Trans. Ind. Appl.*, vol.49, no. 6, pp. 2636–2650, Nov.-Dec. 2013.

3. T. Gopalarathnam and H. A. Toliyat, "A new topology for unipolar brushless DC motor drive with high power factor," *IEEE Trans. Pow. Elect.*, vol. 18, no. 6, pp. 1397–1404, Nov. 2003.

4. P. Alaeinovin and J. Jatskevich, "Filtering of hall-sensor signals for improved operation of brushless DC motors," *IEEE Trans. Energy Convers.*, vol. 27, no. 2, pp. 547–549, Jun. 2012.

5. W. Cui, Y. Gong and M. H. Xu, "A permanent magnet brushless DC motor with bifilar winding for automotive engine cooling application," *IEEE Trans. Magnetics*, vol. 48, no. 11, pp. 3348–3351, Nov. 2012.

6. H. Y. Kanaan and K. Al-Haddad, "A unified approach for the analysis of single-phase power factor correction converters," *37th Annual Conf. on IEEE Ind. Electron. Society (IECON 2011)*, Crown Conference Centre, Melbourne, pp.1167–1172, 7–10 Nov. 2011.

7. C. C. Hwang, P. L. Li, C. T. Liu and C. Chen, C, "Design and analysis of a brushless DC motor for applications in robotics," *IET Elect. Pow. Appl.*, vol. 6, no. 7, pp. 385–389, Aug. 2012.

8. S. B. Ozturk, Oh Yang and H. A. Toliyat, "Power factor correction of direct torque controlled brushless DC motor drive," *42nd IEEE IAS Annual Meeting*, New Orleans, LA, pp. 297–304, 23–27 Sept. 2007.

9. V. Bist and B. Singh, "An adjustable speed PFC bridgeless buck-boost converter fed BLDC motor drive", *IEEE Trans. Ind. Electron.*, vol. 61, no. 6, pp. 2665–2677, Jun. 2014.

10. T. Y. Ho, M. S. Chen, L. H. Yang and W. L. Lin, "The design of a high power factor brushless DC motor drive," *2012 Int. Symposium on Computer, Consumer and Control (IS3C)*, Taichung, pp. 345–348, 4–6 June 2012.

Chapter 5

Optimized feature selection for condition-based monitoring of cylindrical bearing using wavelet transform and ANN

Umang Parmar and Divyang H. Pandya
LDRP Institute of Technology and Research

CONTENTS

5.1 INTRODUCTION

The cylindrical bearings are widely used in the systems that are normally rotating at a lower speed. For finding out the life of a bearing, it becomes essential to understand the failures and their types. The fault diagnosis is one of the methods of understanding the condition of the bearing, and also it helps in the detection of the defects.

As bearings are an internal component of so many big types of machinery and also a very important part from the rotational point of view, it becomes difficult to take it out for the testing purpose. Vibration signals are one of the methods for the detection of faults in the bearing components. Taking signals from both the conditions: healthy and defective, is the preliminary task in this method. Additionally, the application of advanced signal processing method and data mining method for the statistical feature of the signal helps in the perfect identification of the defects and it saves time and increases the life of the machine, too. In the recent industrial revolution, the demand for an auto fault diagnosis method is rising and also it shows accurate results in the recognition of the type of defects, too.

The smooth operation of any machine in the plant is the biggest task for any of the maintenance team. Amplitude, frequency, phase, and modulation

DOI: 10.1201/9781003303053-6

are the four fundamental properties in vibrational research (Babson, 1985). When most of the bearing is rotating at a high speed under the pressure at that time, vibration signals represent the right quality data for the condition monitoring. So many methods such as shock pulse monitoring, analysis of crest factor, a study of kurtosis, high-frequency resonance technique, spectrum analysis, and power spectrum analysis are available for the vibration analysis (McFadden and Smith, 1984).

A new effective tool added in the fault diagnosis area is the optimal wavelet filter selection method. The repetition of the unwanted frequencies highlights the presence of the defects, and with the use of wavelet and statistical parameters such as kurtosis, crest factor, and mean value, the selected envelope of the narrow-band-filtered signal gives the right information, but depending upon the cases such as non-stationary signals that are difficult in the detection of the type of defects (Darji et al., 2020). Denoising of the signals in two ways: first by zero frequency filter and then with the use of wavelet transformation, is used in the early fault detection. It is used for finding out the discontinuity from the vibration signals, and along with that, an optimum wavelet decomposition helps in defining the right frequency band (Sachan et al., 2019).

Based on the similarity of the signals, the concept of mother wavelet is proposed, which gives flexibility in the removal of the noise from the signals. Although it does not apply to all wavelet-based methods, if similarities are there, then it works very effectively with an added advantage of the capability of analyzing stationary as well as non-stationary signals (Rafiee et al., 2010). For the non-stationary signals, the wavelet transfer using the tunable Q-factor also works with the spectral features at the sub-band level. The concept of decomposition at a different level is again proved an important technique (Bharath et al., 2018).

A new hybrid intelligent fault diagnosis and pattern recognition technique based on the EMD and evaluation of distance with wavelet support vector machining gave the result of around 98% in the data set of 200 samples at a very low speed of around 490 RPM (Liu et al., 2013). For the better extraction of the fault data, a dual-tree complex wavelet packet is also found better in the preparation of the feature set from the original signals, and for achieving good results, it also removes the manual feature selection method and adaption of the deep belief network is a replacement of it (Shao et al., 2017).

The use of denoised signals using wavelet transformation instead of the original signals always gives a better performance in the ANN (Kumar et al., 2013). For training and testing of the ANN model with different algorithms, some statistical features from the signals were required to be extracted and the mother wavelet concept helps in that. For the available set of signals, the mother wavelet selection can be done using the minimum Shannon entropy criterion (Kankar et al., 2011).

The present work focuses on the application of mother wavelet selection for denoising the raw signal data and then preparing the data at seventh level decomposition. Here, the selection of features is optimized by reordering them based on the artificial neural network (ANN) performance.

5.2 METHODOLOGY

5.2.1 Advanced signal processing

Conventional signal processing has one limitation in terms of the extraction of some features from the signal, such as breakdown points and discontinuity. To overcome the limitation, advanced signal processing techniques such as wavelet analysis are required. Another advantage of this technique is the extraction of the hidden data using the level decomposition method (Kumar et al., 2013).

When the signal's frequency value is not changing with time, it is called stationary signals. Stationary signals always consist of all frequency components during the entire signal length. Fourier transform (FT) provides the signal's spectral content, but it might fail in providing the information regarding the occurrence timing of that. So, it proves that FT is never going to use when frequency data vary with time. Bearing vibration signals fall under the non-stationary signals; hence, FT is never a good option for analysis. This limitation is then overcome by the introduction of the short-term Fourier transform (STFT). In STFT, it is assumed that the non-stationary signal is stationary in some of its portions, and for that, signals are divided into so many small segments and FT of each segment is generated. Here, the segment of the signals is defined in terms of the width of the segment. If the width is high, it means poor time-related data and reaches frequency data, and if the width is low, the frequency data are flawed compared to the time. Due to the limitation of the reach data of either time or frequency, time-frequency representation is suggested. The wavelet transform is an example of it.

In the time-frequency domain, the signal passes through the two filters: high-pass and low-pass. When this process occurs multiple times, it is called decomposition. For defining the level of decomposition in the present work, the concept of spectral kurtosis is applied, which is defined as:

$$K_Y(f) = \frac{S_{4Y}(f)}{S_{2Y}^2(f)} - 2 \tag{5.1}$$

In the given equation, $y(f)$ stands for the signal and $S_y(f)$ is the classical power spectral density of the signal. After applying this concept, level 7 is defined as the best level of decomposition for the present work (Pandya et al., 2015). Further features are extracted at the level 7 decomposition along with the mother wavelet sym2. This mother wavelet is proved as the best mother wavelet for cylindrical bearings according to the maximum relative wavelet energy criterion and maximum energy-to-Shannon entropy Ratio criteria.

5.2.2 Feature extraction

Extracting the statistical features is one of the key steps in the data mining process. The following features from the denoised signals are extracted and analyzed.

1. Kurtosis is the fourth standardized moment of the signal, where μ stands for the central moment and σ stands for the standard deviation.

$$\text{Kurtosis} = E\left(\frac{X-\mu}{\sigma}\right)^4 = \frac{\mu_4}{\sigma^4} \tag{5.2}$$

2. The energy of the signal is defined as the total area that is under the magnitude's square of the signal $x(t)$.

$$\text{Energy} = \int_{-\infty}^{\infty} |x(t)|^2 \, dt \tag{5.3}$$

3. The mean of the signal is defined by the below equation for the signal $x(t)$.

$$\text{Mean}(\mu) = \frac{1}{N}\sum_{i-1}^{N-1} X_i \tag{5.4}$$

4. The standard deviation σ is the average deviation that is not using the amplitude, but the power.

$$\sigma = \left[\frac{1}{N-1}\sum_{i-1}^{N-1}(X_i - \mu)^2\right]^{1/2} \tag{5.5}$$

5. Skewness is defined as the amount of the symmetricity around the mean value, and the formula of it is as below.

$$\text{Skewness} = \frac{E(X-\mu)^3}{\sigma^3} \tag{5.6}$$

6. The crest factor is defined as the ratio of the peak to RMS value of the waveform.

$$\text{Crestfactor} = \frac{|X_{\text{peak}}|}{X_{\text{RMS}}} \tag{5.7}$$

7. The mean square value is the root of the arithmetic mean of the sum of the square values.

$$\text{MeanSquare} = \lim_{T \to 0} \sqrt{\frac{1}{T} \int_0^T |X(t)| dt} \tag{5.8}$$

The above features are some of the features selected as the statistical parameters for training the data in the ANN. Other features such as the maximum value from the signal are also calculated for all the signals. It is noted that all these parameters have their own mathematical range so that the normalization of the data is required before being applied to the ANN. All data are normalized between the 0 and 1 values for the ease of algorithm training.

5.3 DATA ACQUISITION

5.3.1 Data preparation

The experimental setup consists of basic elements such as motor, shaft, bearing, housing, COCO analyzer, and sensor cables, as shown in Figure 5.1. Here, a solid platform is created for the setup and a metal plate is bolted on the top surface of the platform. Over the plate, one rigid clamping and one flexible metal clamping are bolted. The flexible clamp provides the flexibility of putting shaft with some variation of length. The shaft is mounted on the setup with the use of a cylindrical roller bearing for taking the vibration signals.

Vibration signals are taken with the use of the sensor cables attached to the analyzer. Signals are taken at various speeds, and the speed is controlled by the speed controller attached with the DC servo motor. Four set sets of bearings are used to have a healthy bearing, bearing with inner race, bearing with outer race, and bearing with roller defects. Defects are of size 1.5 mm made using the laser cutting operation as shown in Figure 5.2. The specification of the bearing is shown in Table 5.1.

Figure 5.1 Experimental setup.

Figure 5.2 Defects on the inner race, outer race, and on the roller.

5.3.2 Data mining

In the recent past, the preparation of a system that can think like a human has become a trend. But there is a difference between human thinking and machine behavior; humans understand the pattern in the data, whereas the machine just recognizes. Humans generally make a pattern in the mind using the neural network and store the data in terms of the biological sensors available in the body, while an ANN also acts the same as the human mind and, according to the algorithm, it trains the data, finds the pattern, and test the data for the prepared pattern (Yegnanarayana, 1994).

Table 5.1 Details of the bearing

Parameters	Details
Bearing	NJ305
Type of the bearing	Cylindrical roller bearing
Inside diameter	25 mm
Outside diameter	62 mm
Width	17 mm
Weight	0.24 kg
Number of raceway ring rows	Signal row
Number of rollers	10
Radial clearance	20–45 μm

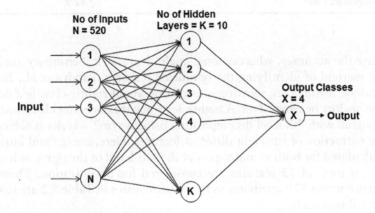

Figure 5.3 ANN network.

Here, in this research, an ANN is trained with nnstart toolbox in MATLAB. Pattern recognition networks are the feedforward networks for the training and the classification of the input data into targets. The targets often known as classes are defined as the condition of the bearing, so here four target values are considered in terms of the 900 (sample)×4 (classes) matrix. The hidden size is taken as ten, which is the default value. The total input data are of a 900 (sample)×520 (features) matrix as shown in Figure 5.3. Out of the 900-sample data, 70% (630) of the data are used for training purpose and the other 15%–15% (135 samples each) data are kept for validation and testing.

5.4 RESULTS AND DISCUSSION

In the training of the ANN model, the total features selected play a major role. All features are not useful in the ANN model training. Some can

Table 5.2 Each feature's % accuracy for the training of the ANN model

Feature name	% Accuracy	Rank
Kurtosis x-axis (level 7 – 128 attributes)	41.7	7
Kurtosis y-axis (level 7 – 128 attributes)	62.4	3
Energy x-axis (level 7 – 128 attributes)	71.8	2
Energy y-axis (level 7 – 128 attributes)	82.3	1
Mean value	32.6	12
Std. deviation	47.4	4
Energy	45.7	5
Skewness	39.6	9
Kurtosis of signal	34.4	10
Crest factor	32.7	11
Max. value	40.9	8
Mean-squared value	42.9	6

increase the accuracy, whereas some might decrease it. Sensitivity analysis is one method of identifying the optimized features (Pandya et al., 2013). All trained features are showing information about rightly classified defective or healthy bearings data. A higher accuracy means better classification. Each signal with 7th level decomposition using a sym2 wavelet is subjected to the extraction of the eight different features. Here, energy and kurtosis are calculated for both x- and y-axes at the sub-band of the signal at level 7. Hence, a total of 12 features are considered for optimization. These 12 features contain 520 attributes as first four features in Table 5.2 are having 128 attributes each.

In this research, a total of 900 signals (225 of each condition) were taken for the training purpose. Data taken at different RPMs range from 200 to 3,000 (Yadav and Pandya, 2017). Twelve features of all signals in terms of the numerical data show the detailed information of the condition. Here, data are normalized and then each feature individually is trained in the model. Features are ranked according to their accuracy in the ANN model in descending order.

Finally, the model is trained with the first-ranked feature and after training, next ranked feature is added for calculation of the model accuracy. By this way, the feature ranking technique helps in the evaluation of the model's sensitivity to feature selection. Optimized numbers of features are noted as 10 out of the 12 selected features, with an accuracy of 97.6% as shown in Figure 5.4.

The overall process of auto fault detection is as below:

- Signals from the cylindrical bearing at four different conditions are denoised with the best mother wavelet at decomposition level 7.

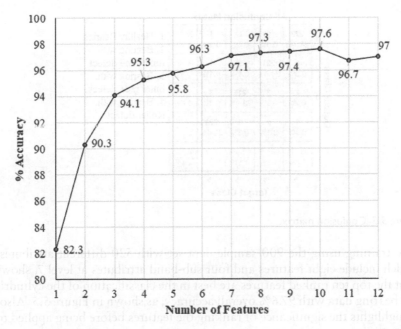

Figure 5.4 Number of features vs % accuracy.

- Twelve features from 900 sample signals are extracted.
- Data normalization is done for removing the effect of the numerical range.
- Every feature is ranked according to its individual performance in ANN.
- A combination of all features is tested, and an optimum number of features and their combination is selected for training, validation, and model testing for fault detection.

5.5 CONCLUSIONS

In this chapter, a cylindrical bearing fault diagnosis is explained using the wavelet analysis and ANN technique. A similar work on the ball bearing with lesser features reported approx. 99% accuracy in training and also proved better than the KNN technique (Gunerkar et al., 2019), while another stated an accuracy of 96%, which was better than the support vector machine technique (Agrawal and Jayaswal, 2020). Another work on the ball bearing also showed a result of approx. 97% with sub-band kurtosis and energy as features (Pandya et al., 2012). Ball bearing result comparison for ANN and SVM in one more research showed that ANN was better than SVM (Darji et al., 2019).

Here, along with feature extraction, additionally feature ranking is added in terms of improving the accuracy and avoiding unnecessary data training.

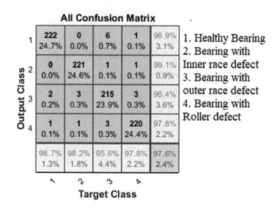

Figure 5.5 Confusion matrix.

The training using the 900-sample data set with 520 different attributes, which include eight features and four sub-band attributes at level 7, shows that the top ten ranked features are best in the classification of the cylindrical bearing data with 97.6% overall accuracy, as shown in Figure 5.5. Also, it highlights the significance of ranking the features before being applied to data mining.

REFERENCES

Agrawal, P. and Jayaswal, P. (2020). Diagnosis and classifications of bearing faults using artificial neural network and support vector machine. *Journal of The Institution of Engineers (India): Series C*, 101(1), 61–72.

Babson, J.T. (1985). Vibration analysis-a proven technique as a predictive maintenance tool. *IEEE Transactions on in Industry Applications*, IA(2), 1–21.

Bharath, I., Devendiran, S. and Mathew, A. T. (2018). Bearing condition monitoring using tunable Q-factor wavelet transform, spectral features and classification algorithm. *ICMMM, Materials Today: Proceedings*, 5(5), 11476–11490.

Darji, A., Darji, P. H. and Pandya, D. H. (2019). Fault diagnosis of ball bearing with WPT and supervised machine learning techniques. *Advances in Intelligent Systems and Computing*, 748, 291–301.

Darji, A. A., Darji, P. H. and Pandya, D. H. (2020). Envelope spectrum analysis with modified EMD for fault diagnosis of rolling element bearing. *1ST International Conference & 4TH National Conference on Reliability and Safety Engineering (INCRS -2018)* (pp. 91–100). Jabalpur, Springer.

Gunerkar, R. S., Jalan, A. K. and Belgamwar, S. U. (2019). Fault diagnosis of rolling element bearing based on artificial neural network. *Journal of Mechanical Science and Technology*, 33(2), 505–511.

Kankar, P. K., Sharma, S. C. and Harsha, S. P. (2011). Rolling element bearing fault diagnosis using wavelet transform. *Neurocomputing*, 74(10), 1638–1645.

Kumar, H. S., Pai, P. S., Sriram, N. S. and Vijay, G. S. (2013). ANN based evaluation of performance of wavelet transform for condition monitoring of rolling element bearing. *International Conference on Design and Manufacturing* (pp. 805–814). Chennai, Procedia Engineering, Elsevier.

Liu, Z., Cao, H., Chen, X., He, Z. and Shen, Z. (2013). Multi-fault classification based on wavelet SVM with PSO algorithm to analyze vibration signals from rolling element bearings. *Neurocomputing*, 99, 399–410.

McFadden, P.D. and Smith, J.D. (1984). Vibration monitoring of rolling element bearings by the high frequency resonance technique a review. *Tribology International*, 17(1), 3–10.

Pandya, D. H., Upadhyay, S. H. and Harsha, S. P. (2012). ANN based fault diagnosis of rolling element bearing using time-frequency domain feature. *International Journal of Engineering Science and Technology*, 4(6), 2878–2886.

Pandya, D. H., Upadhyay, S. H. and Harsha, S. P. (2013). Fault diagnosis of rolling element bearing with intrinsic mode function of acoustic emission data using APF-KNN. *Expert Systems with Applications*, 40(10), 4137–4145.

Pandya, D. H., Upadhyay, S. H. and Harsha, S. P. (2015). Fault diagnosis of high-speed rolling element bearings using wavelet packet transform. *International Journal of Signal and Imaging Systems Engineering*, 8(6), 390–401.

Rafiee, J., Rafiee, M. A. and Tse, P.W. (2010). Application of mother wavelet functions for automatic gear and bearing fault diagnosis. *Expert Systems with Applications*, 37(6), 4568–4579.

Sachan, S., Shukla, S. and Singh, S.K. (2019). Two level de-noising algorithm for early detection of bearing fault using wavelet transform and zero frequency filte. *Tribology International*, 143, 1–27.

Shao, H., Jiang, H., Wang, F. and Wang, Y. (2017). Rolling bearing fault diagnosis using adaptive deep belief network with dual-tree complex wavelet packet. *ISA Transactions*, 69, 187–201.

Yadav, H. K. and Pandya, D. H. (2017). Non linear dynamic analysis of cylindrical roller bearing. *11th International Symposium on Plasticity and Impact Mechanics* (pp. 1878–1885). New Delhi, Procedia Engineering.

Yegnanarayana, B. (1994, April 12). Artificial neural networks for pattern recognition. *Sadhana*, 19(2), 189–238.

Chapter 6

SafeShop – an integrated system for safe pickup of items during COVID-19

Nisarg Kapkar, Jahnavi Shah, Sanyam Raina,
Nisarg Dave, and Mustafa Africawala
Pandit Deendayal Energy University

CONTENTS

6.1 INTRODUCTION

It's been 2 years since [1] COVID-19 (also known as coronavirus) took a toll on the world. According to [2] WHO's COVID-19 dashboard, over 340 million people were infected and 5.5 million people died because of COVID-19 in 2 years. [3] COVID-19 mainly spreads when an infected person is in close contact with another person. While there is no effective cure found for the same, vaccines, safety measures, and precautions are taken to ensure the containment of the virus and to slow the spread of the virus. Getting vaccinated, [4] wearing a mask, and [5] ensuring social distancing have become of utmost importance. Due to the lack of ensuring a safe method to reopen, businesses have suffered a lot. SafeShop is an integrated system made on two of the main pillars of containing the virus – social distancing and wearing a mask. The camera feed at the entrance of the shops takes a snapshot of the person entering the shop, which is then passed to the backend of our machine learning algorithm that permits the person to

DOI: 10.1201/9781003303053-7

enter if they are wearing the mask correctly. The location of everyone in the enclosed space is constantly fetched, and the distance is calculated based on it and displayed as a graph to ensure proper social distancing. We have achieved an accuracy of 94.67% in our face mask detection model, and our social distancing algorithm constantly fetches the location to ensure safety all the time. SafeShop ensures that all the shops can reopen safely, and the people visiting these shops also feel safe. We exploit the recent developments in technologies to build a fully integrated system.

6.2 LITERATURE SURVEY

6.2.1 Face mask detector

The goal is to recognize the people who are not wearing a mask or are wearing the mask incorrectly to help decrease the spread of COVID-19. Face mask detection can be solved by traditional object detection algorithms such as Haar cascades, but these methods require face engineering. Now, it is possible to train neural networks that do not require feature engineering and can provide better results than traditional algorithms.

In paper [6], the authors used a combination of SR network with a classification network (SRCNet) to identify three facemask-wearing conditions: correct facemask-wearing, incorrect facemask-wearing, and no facemask-wearing.

In paper [7], the proposed model for face mask detection included two components: The first component is deep transferring learning (ResNet50) as a feature extractor and the second component is classification machine learning (support vector machine).

In paper [8], the proposed model consisted of two stages: The first stage included a face detector and the second stage is a CNN-based face mask classifier.

6.2.2 Mobile application and social distance algorithm

The goal is to create a mobile application that will allow people to place orders and a distance tracking algorithm that tracks the distance between nearby users and notifies users if there are people within their 6 feet range.

In paper [9], the mobile application uses live location through GPS to send the location to specific contact numbers. Inspired by this, we have implemented live location tracking through GPS in our mobile application and the coordinates (latitude and longitude) are shown on the location page. Furthermore, the application will also display a live graph showing nearby users and their relative position (distance and angle) with respect to one another.

In paper [10], the mobile application allows users to order food and keep its record. Each restaurant has its order page. Inspired by this, we have implemented an online food ordering system where each restaurant has its menu page from which users could choose items to order. Also, we implemented a cart page that keeps a record of our order and gives a summary of the order before proceeding to pay for it.

In paper [11], the mobile application allows users to place their orders and gives them a pickup time at a specific merchant location. It also includes menu pages for restaurants, prices, etc. Inspired by this, we have implemented this feature in our mobile application, where after placing an order, users will get a pickup time for their order along with various other information. Also, we have implemented a menu page for each restaurant along with the price of various items.

6.3 METHODOLOGY

6.3.1 Face mask detector

The face mask detection model is built using TensorFlow 2 Object Detection API on Google Colaboratory. The SSD (single-shot detector) ResNet50 V1 FPN 640×640 has been used as a pre-trained network. The face mask detection model is built on top of the pre-trained network.

The ssd_resnet50_v1_fpn model [12] is SSD FPN object detection architecture based on ResNet50. The model has been trained from the Common Objects in Context (COCO) image dataset.

The SSD [13] has two components: a backend model and a SSD head.

The backend model usually is a pre-trained image classification network (in this case, ResNet50). The SSD head is just one or more convolution layers added to the backbone. The below flowchart (Figure 6.1) shows the proposed deep learning architecture.

The face mask detection model is classified under supervised learning. In supervised learning, the computer is trained on data containing both input

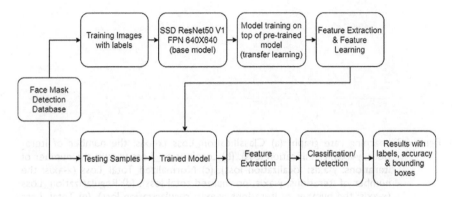

Figure 6.1 Proposed deep learning architecture.

(images) and output (labels), and the goal is to learn patterns and predict which class an object belongs to.

The model is trained on more than 1,000 faces taken from the Kaggle database [14]. An additional 500 images were scrapped from the Internet and used for training the model. The model was trained until the total loss was constantly less than 0.4.

Figure 6.2 shows the change in learning rate during model training, and Figure 6.3 shows the change in various losses during model training.

The model can identify three classes (Figure 6.4):

- with_mask: The person is wearing a mask, and their nose and mouth are properly covered.
- without_mask: The person is not wearing a mask.
- mask_weared_incorrect: The person is wearing a mask, but their nose/mouth are not covered properly.

6.3.2 Mobile application

The prototyping and designing of pages of the mobile application were done in Adobe XD. The mobile application was built on Android Studio using the languages Java (for backend) and XML (for designing and structuring of the pages).

Each page of the app, as shown in Figure 6.5, has a bottom navigation bar which allows you to quickly access the Home page, Cart page, and Settings page.

The Home page of the app will show you various registered restaurants and also the restaurants near you. It also provides you options to switch to grocery shops if you wish to buy groceries and also medical shops if you wish to buy medical products.

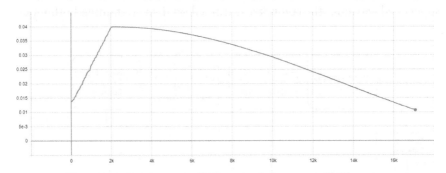

Figure 6.2 Learning rate graph. (a) Classification_Loss (x-axis: the number of iterations; y-axis: classification loss). (b) Localization_Loss (x-axis: the number of
iterations; y-axis: localization loss). (c) Normalized_Total_Loss (x-axis: the
number of iterations; y-axis: normalized total loss). (d) Regularization_Loss
(x-axis: the number of iterations; y-axis: regularization loss). (e) Total_Loss
(x-axis: the number of iterations; y-axis: total loss).

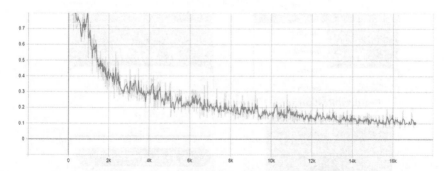

(a) classification_loss (x-axis: number of iterations; y-axis: classification loss)

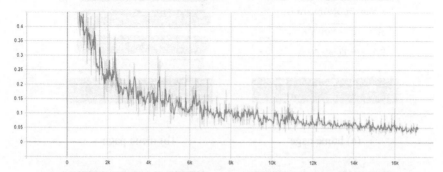

(b) Localization_Loss (x-axis: number of iterations; y-axis: localization loss)

Figure 6.3 Loss graph.

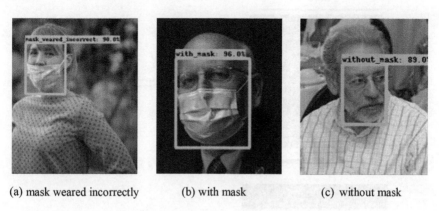

(a) mask weared incorrectly   (b) with mask   (c) without mask

Figure 6.4 Trained face mask detector model predictions.

On clicking any of the restaurants, you will be directed to their respective menu page, where you could place your order by selecting how many of the given items you want by updating the counter.

(a) home page

(b) menu page

(c) cart page

(d) current location page

Figure 6.5 Mobile application prototype.

Once you update the counters as per your wish, you could click on the "Add To Cart" button to add your order to the cart where you could verify and review your order. The Cart page will show you the order you added. It will also show you your Order ID, pickup time, and the total amount to be paid along the address of the respective restaurant.

When the user is within 10 feet range of the pickup location, their live location will be sent to Google Firebase real-time database (see Algorithm 1 below).

Their live location will be updated every few seconds. Similarly, this is done for all the active users within the 10 feet range of the pickup location. Based on the live location of all users, a graph, as shown in Figure 6.6, is made, which shows the distance between an individual user and all the nearby users. A blue dot represents the individual user, and all other nearby users are represented by either a red dot or a green dot depending on the distance.

- Red dot: That person is within 6 feet range of the user.
- Green dot: That person is not within 6 feet range of the user

Users can view the graph and maintain appropriate distance based on that graph. Individual users will also be notified when someone enters their 6 feet range.

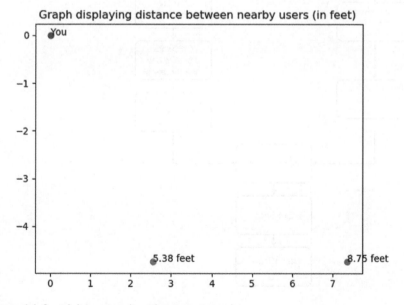

Figure 6.6 Social distancing algorithm output graph.

Algorithm 1: To calculate the distance and angle between people

Input: the location of the user
while current user ≤ 10 feet of shop: **do**
Find the location (latitude and longitude) of current_user
    Google Firebase ← location of current_user
    for nearby_user **do**
        Location ← Google Firebase
        Calculate distance between current_user and nearby_user

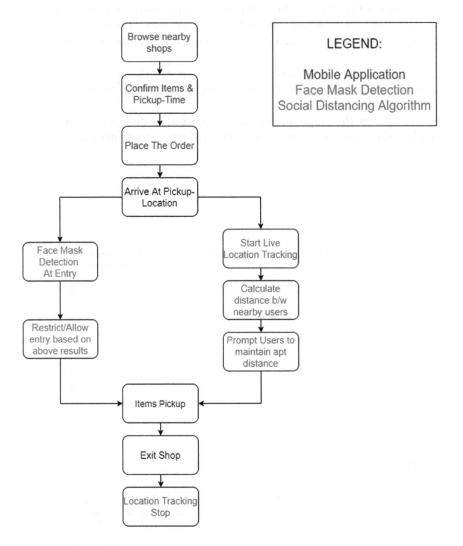

Figure 6.7 Overall flowchart.

    Calculate relative angle between current_user and
nearby_user

    Graph ← Calculated angle and distance

  end

  Output: a graph showing the distance between the current_user
and nearby_users

  end

The above flowchart (Figure 6.7) shows the overall approach of our
solution.

6.4 RESULTS AND DISCUSSION

6.4.1 Face mask detector

The architecture of our system is to check whether the person entering the
shop is wearing a mask properly or not. This captures your face with a camera near the entrance of shops and analyzes how the mask is worn.

It is then divided into three classes: with_mask, without_mask, and
mask_weared_incorrect, and we get a live response on the screen where the
person can see the error. These cameras are connected to a system including
a speaker, display, and auto-door. These will notify the person and prompt
them to wear a mask properly before entering the shop. We tested the model
on around 150 images scraped from the Internet. We achieved an accuracy
of 94.67% in our face mask detection model.

6.4.2 Mobile application and social
   distancing algorithm

The app will allow people to order and collect items safely. This will reduce
the exposure to the virus by converting physical waiting lines into digital
waiting queues.

Meanwhile, the social distance tracking algorithm will send the location (latitude and longitude) to Google Firebase real-time database, tracking the movements of people, calculating the distance using that location from Firebase,
and informing the user whether they are standing too close (less than 6 feet
distance) with a live graph display, thereby maintaining social distancing.

The live location (latitude and longitude) is precise up to seven decimal places [15]. Seven decimal implies that the location is accurate up to
approximately 1.11 cm.

Potential factors for inefficiency in location tracking:

- [16] Error and mismatch in the location might occur if the user is
  near a building or a tree. The location tracking also depends on the
  signal strength and atmospheric conditions. Poor signal strength

and improper atmospheric conditions may cause inaccurate location tracking.

- The location tracking system is limited by hardware. Older devices might provide more inaccurate results (compared to newer devices).

Few inaccuracies might also arise in social distancing algorithms:

- Slight precision error in distance and angle calculations might occur because of rounding error and usage of float/double data types.
- Accuracy error from live location tracking might carry over, resulting in inaccurate calculation of distance and algorithm.

However, these errors are quite negligible when compared to 6 feet social distancing.

6.5 CONCLUSIONS

This chapter highlights the working of face mask detection, a mobile application with the social distance tracking algorithm in shops/restaurants. The purpose of this paper is to discuss the methodology adopted for face mask detection, a mobile app with the distance tracking algorithm for efficient social distancing.

During the research, thorough research and study were conducted on the existing system and proposed system with all its technologies, which have aptly been portrayed in the form of a literature review.

Future plans include

- Accuracy error from live location tracking might carry over, resulting in inaccurate calculation of distance and algorithm.
- For social distancing algorithm, combining the algorithm with the mobile application.

ACKNOWLEDGEMENTS

We would like to express our deep gratitude to Kaustubh Sadekar for his valuable and constructive suggestions during the development of this research work.

REFERENCES

1. Covid-19 Wikipedia. URL https://en.wikipedia.org/wiki/COVID-19.
2. Who coronavirus disease (covid-19) dashboard. URL https://covid19.who. int/.
3. Coronavirus disease (COVID-19): How is it transmitted? URL https://www.who.int/news-room/questions-and-answers/item/coronavirus-disease-covid-19-how-is-it-transmitted.
4. How well do face masks protect against COVID-19? URL https://www.mayoclinic.org/diseases-conditions/coronavirus/in-depth/coronavirus-mask/art-20485449.
5. M. Greenstone, and V. Nigam. (2020). Doi: 10.2139/ssrn.3561244.
6. B. Qin, and D. Li. Identifying facemask-wearing condition using image super resolution with classification network to prevent COVID-19. *Sensors*, 20(18), 5236, (2020).
7. M. Loey, G. Manogaran, M.H.N. Taha, and N.E.M. Khalifa. A hybrid deep transfer learning model with machine learning methods for face mask detection in the era of the covid-19 pandemic. *Measurement*, 167, 108288, (2021).
8. A. Chavda, J. Dsouza, S. Badgujar, and A. Damani, Multi-Stage CNN architecture for face mask detection (2020). https://arxiv.org/pdf/2009.07627
9. R.S. Yarrabothu, and B. Thota. *Annual IEEE India Conference (INDICON)*, pp. 1–4 (2015).
10. K. Bhandge, T. Shinde, D. Ingale, N. Solanki, and R. Totare. A proposed system for touchpad based food ordering system using android application. *International Journal of Advanced Research in Computer Science Technology (IJARCST 2015)* 3, 70–72, (2015).
11. S. Elston, and B. Smith, Patent application publication (2002).
12. Open vino documentation for SSD resnet50 v1 FPN coco. URL https://docs.openvino.ai/latest/omz_models_model_ssd_resnet50_v1_fpn_coco.html.
13. How single-shot detector (SSD) works? URL https://developers.arcgis.com/python/guide/how-ssd-works/#:~:text=SSD%20has%20two%20components%3A%20a, classification%20layer%20has%20been%20removed.
14. Mask dataset. URL https://makeml.app/datasets/mask.
15. Decimal degrees. URL http://wiki.gis.com/wiki/index.php/Decimal_degrees.
16. GPS accuracy. URL https://www.gps.gov/systems/gps/performance/accuracy/#how-accurate.

Chapter 7

Solution to first-order fuzzy differential equation using numerical methods

Meghna Parikh, Manoj Sahni, and Ritu Sahni

Pandit Deendayal Energy University

CONTENTS

7.1 INTRODUCTION

We come across various situations such as imprecise, ambiguous, confusing, inadequate, and inconsistent circumstances in our day-to-day lives. It becomes difficult to accurately model and address problems with limited data. Fuzzy set theory, a classical set theory generalization, describes situations where the data are imprecise or ambiguous. To handle such situations,

DOI: 10.1201/9781003303053-8

79

in 1965, Zadeh [1] developed fuzzy set theory as an extension of classical set theory. In real-world situations, we consistently get info in the form of ambiguous phrases such as "small," "very small," "long," and "very long"; all these facts might differ for every person since they are linked to human mind and hence vulnerable to human aspects. In fuzzy set theory, these concepts are called "linguistic words," and their significance determines their membership grade. Fuzzy sets are made by using these linguistic words. Together with their membership degrees, all these linguistic notions are represented as ordered pairs and then utilized to build fuzzy sets. These fuzzy values correspond to the grade of membership in a fuzzy set A, which is well defined over the universal set X and has the grade of membership $\mu_A(x) \in [0,1]$.

The intuitionistic fuzzy set (IFS), initially expressed by Atanassov in 1986 [2], is a further extension of fuzzy sets, where it is considered membership and non-membership grade. The fuzzy sets are defined by observing the workings that only study the grade of membership of any information and not the grade of non-membership. In some cases, evaluating only the membership value is unsatisfactory; the non-membership value must also be considered.

In real life, differential equations develop mathematical models in many fields, including science, engineering, physics, and medicine. As a result, working on developing to solve differential equations is vital for proceeding in any area. Nowadays, fuzzy logic has emerged as an essential and helpful tool for modeling numerous real-world phenomena using differential equations with unknown and imprecise parameters in all fields of science and engineering. Dubois et al. [3–5] introduced the application of fuzzy differential equation (FDE) by focusing solely on membership values. Chang et al. [6] discovered fuzzy numbers and fuzzy functions for defining FDE.

Furthermore, the concept of intuitionistic fuzzy differential equations [7,8] was developed, which contained both membership and non-membership values as parameters. It is vital to understand the concept of derivatives in a fuzzy environment to examine the solutions to these fuzzy differential equations. Much work has been published in the literature on this topic, such as differentials for fuzzy functions investigated by Puri and Ralescu [9] and fuzzy calculus explored by Goetschel and Voxman [10]. Researchers [11–15] employed the fuzzy derivative concept to solve first-order ordinary differential equations with initial conditions. They also demonstrated how to use the lower-upper form of fuzzy integers to solve fuzzy differential equations with this generalized differentiability. Stefanini et al. [16] used the generalized Hukuhara derivative to illustrate the generalization of the fuzzy interval-valued function in 2009.

If we discuss fuzzy differential equations, intuitionistic fuzzy differential equations are more generalized since they consider membership and non-membership values. The review shows that various analytical, semi-analytical, and numerical methodologies have been formed and used to find acceptable solutions to fuzzy differential equations. For example,

Salahshour et al. [17] developed the Laplace transform for a fuzzy differential equation, and Mondal et al. [18,19] utilized it to solve fuzzy differential equations. In 2016, Tapaswini et al. [20] solved a fuzzy differential equation using an analytical approach. Ahmadian et al. [21] utilized the fourth-order Runge-Kutta method to solve a fuzzified differential equation. Similarly, Sahni et al. [22–24] and Parikh et al. employed an analytical method to solve a second-order fuzzy differential equation in an uncertain environment. In this chapter, we have used the numerical techniques to solve an ordinary differential equation of first order in the both fuzzy and intuitionistic environments. The solutions are obtained at different α-cut and (α, β)-cut points and time values, respectively. The results are compared with the solution obtained by taking the crisp set values at different times.

7.2 PRELIMINARIES

Here, we discuss some required preliminaries as follows.

7.2.1 H-differentiability

A function $f : (a, b) \rightarrow R_F$ is called H-differentiable on $x_0 \in (a,b)$ if for $h > 0$ sufficiently small, there exist the H-differences $f(x_0 + h)\ominus_H f(x_0)$, $f(x_0)\ominus_H f(x_0 - h)$ and an element $f'(x_0) \in R_F$ such that [15]

$$\lim_{h \to 0} \frac{f(x_0 + h)\ominus_H f(x_0)}{h} = \lim_{h \to 0} \frac{f(x_0)\ominus_H f(x_0 - h)}{h} = f'(x_0)$$

7.2.2 Strongly generalized differentiability

Let $f : (a, b) \rightarrow R_F$ and $x_0 \in (a,b)$; we say that f is strongly generalized differentiable if there exists an element $f''(x_0) \in E$ such that [15]

i. for all $h > 0$ sufficiently small exists, $f'(x_0 + h)\ominus_H f'(x_0), f'(x_0)\ominus_H f'(x_0 - h)$ and the limits hold (in the metric D)

$$\lim_{h \to 0} \frac{f'(x_0 + h)\ominus_H f'(x_0)}{h} = \lim_{h \to 0} \frac{f'(x_0)\ominus_H f'(x_0 - h)}{h} = f''(x_0) \text{ or } \quad (7.1)$$

ii. for all $h > 0$ sufficiently small exists, $f'(x_0)\ominus_H f'(x_0 + h), f'(x_0 - h)\ominus_H f'(x_0)$ and the limits hold (in the metric D)

$$\lim_{h \to 0} \frac{f'(x_0)\ominus_H f'(x_0 + h)}{h} = \lim_{h \to 0} \frac{f'(x_0 - h)\ominus_H f'(x_0)}{h} = f''(x_0) \text{ or } \quad (7.2)$$

iii. for   all $h > 0$ sufficiently   small   exists, $f'(x_0 + h)\theta_H f'(x_0), f'(x_0 - h)\theta_H f'(x_0)$ and the limits hold (in the metric D)

$$\lim_{h \to 0} \frac{f'(x_0 + h)\theta_H f'(x_0)}{h} = \lim_{h \to 0} \frac{f'(x_0 - h)\theta_H f'(x_0)}{h} = f'(x_0) \text{ or } \quad (7.3)$$

iv. for   all $h > 0$ sufficiently   small   exists, $f'(x_0)\theta_H f'(x_0 + h), f'(x_0)\theta_H f'(x_0 - h)$ and the limits hold (in the metric D)

$$\lim_{h \to 0} \frac{f'(x_0)\theta_H f'(x_0 + h)}{h} = \lim_{h \to 0} \frac{f'(x_0)\theta_H f'(x_0 - h)}{h} = f''(x_0) \quad (7.4)$$

7.2.3 Fuzzy set (FS)

Let X be any universal set and $\mu_A(x) \in [0,1], \forall x \in X$ be a membership value defined for each element x in the universal set X. Then the fuzzy set A is defined as $A = \{(x, \mu_A(x)) \mid \forall x \in X\}$.

7.2.4 α-Level of fuzzy set

The α-level of fuzzy set A is defined as $A_\alpha = \{\mu_A(x) \geq \alpha, \alpha \in [0, 1]\}$, where $x \in X$. This set includes all the elements of X with membership values in A that are greater than or equal to α [6].

7.2.5 Triangular fuzzy number

If A is a fuzzy set, the triangular fuzzy numbers are defined as three different points of $A = (a_1, a_2, a_3)$, which is represented as the membership function of $\mu_A(x)$ [7].

$$\mu_A(x) = \begin{cases} \dfrac{x - a_1}{a_2 - a_1} & \text{for } a_1 \leq x < a_2 \\ 1 & x = a_2 \\ \dfrac{a_3 - x}{a_3 - a_2} & \text{for } a_2 \leq x < a_3 \\ 0 & \text{otherwise} \end{cases}$$

7.2.6 Intuitionistic fuzzy set (IFS)

An intuitionistic fuzzy set B over universal set of X is represented by $B = \{(x, \mu_B(x), \upsilon_B(x)) \mid \forall x \in X\}$, where $\mu_B(x)$ represents the membership value of x in B, and value $v_B(x)$ represents the non-membership value of x in B [6].

7.2.7 α, β-Level of intuitionistic fuzzy set

An $\alpha,$ β-level of intuitionistic fuzzy set B is defined as $B_{\alpha,\beta} = x : \{\mu_B(x) \geq \alpha, v_B(x) \leq \beta, \forall x \in X, \alpha, \beta \in [0,1]\}$ with $\alpha + \beta \leq 1$, and X is the universal set [6]

7.2.8 Triangular intuitionistic fuzzy set

If A is fuzzy set, the triangular fuzzy numbers are defined as three different points of $A = (a_1, a_2, a_3)$, which is represented as the membership function of $\mu_A(x)$ and non-membership function $\vartheta_N(x)$ [6]:

$$\mu_M(x) = \begin{cases} \left(\dfrac{x-a}{b-a}\right), & \text{for } a \leq x \leq b \\ 1 & \text{for } x = b \\ \left(\dfrac{c-x}{c-b}\right), & \text{for } b \leq x \leq c \\ 0 & \text{otherwise} \end{cases}$$

$$\vartheta_N(x) = \begin{cases} \left(\dfrac{b-x}{b-a}\right), & \text{for } a \leq x \leq b \\ 0, & \text{for } x = b \\ \left(\dfrac{x-c}{c-b}\right) & \text{for } b \leq x \leq c \\ 1 & \text{otherwise} \end{cases}$$

7.3 METHODOLOGY

7.3.1 Solution to first-order differential equation in fuzzy environment using Euler's method

Let us consider the first-order differential equation as follows:

$$\frac{dy}{dt} = f(t, y(t)), \quad y(t_0) = y_0 \tag{7.5}$$

where y is a dependent variable, t is an independent variable, and t_0 is the initial value of the parameter t. Here, we consider initial values in the fuzzy environment defined as:

$$y(t_0) = \left[a + \alpha(b-a), c - \alpha(c-b) \right] \qquad (7.6)$$

where $y(t_0)$ represents the membership grade in the fuzzy environment.

Considering α-cut definition for the given differential equation, the modified differential equation (7.5) in the fuzzy environment can be written as: $\dfrac{dy}{dt} = f\left(t_n, \left[\left(\underline{y}(t,\alpha), \overline{y}(t,\alpha) \right) \right] \right)$ with the initial condition $y(t_0)_\alpha = \left[\underline{y_T}(t_0,\alpha), \overline{y_T}(t_0,\alpha) \right]$.

Using Taylor's theorem, we develop Euler's method in the fuzzy environment. Suppose that $y(t)$ is the approximate solution to the given equation in the interval $[t_0, P]$; then using Taylor's theorem, we have

$$y(t_{n+1}) = y(t_n) + (t_{n+1} - t_n)y'(t_n) + \frac{1}{2}(t_{n+1} - t_n)^2 y''(\zeta) + \cdots \qquad (7.7)$$

Here, for each $n = 0,1,2,\ldots, N - 1$, equation (7.7) holds well for some numbers $\zeta_n \in (t_n, t_{n+1})$. Taking $h = t_{n+1} - t_n$ in equation (7.7), we get

$$y(t_{n+1}) = y(t_n) + hy'(t_n) + \frac{1}{2}h^2 y''(\zeta) + \cdots \qquad (7.8)$$

Since $y'(t_n) = f(t_n, y(t_n))$ is an approximate solution to the given differential equation (7.5), we can write it as

$$y(t_{n+1}) = y(t_n) + hf(t_n, y(t_n)) + \frac{1}{2}h^2 y''(\zeta) + \cdots \qquad (7.9)$$

By truncating the higher-order terms from second order in equation (7.8) and denoting $y(t_0) = \left[\underline{y}(t), \overline{y}(t) \right]$, we obtain the following equations

$$\underline{y}(t_{n+1}) = \underline{y}(t_n) + hf\left(t_n, \underline{y}(t_n) \right) \qquad (7.10)$$

$$\overline{y}(t_{n+1}) = \overline{y}(t_n) + hf\left(t_n, \overline{y}(t_n) \right) \qquad (7.11)$$

Here, $y(t_{n+1}) = \left[\underline{y}(t_{n+1}), \overline{y}(t_{n+1}) \right]$ represents the solution in the form of lower and upper bounds of membership at time t_{n+1}. Equations (7.10) and (7.11) represent Euler's method in the fuzzy environment.

7.3.2 Solution to first-order differential equation in intuitionistic fuzzy environment using Euler's method

Let the given ordinary differential equation of first order be as follows:

$$\frac{dy}{dt} = f(t, y(t)), \quad y(t_0) = y_0 \tag{7.12}$$

where y is a dependent variable, t is an independent variable, and t_0 is the initial value of the parameter t. Here, we consider initial values in the intuitionistic fuzzy environment defined as:

$$y_M(t_0) = \left[a + \alpha(b - a), c - \alpha(c - b) \right] \tag{7.13}$$

$$y_N(t_0) = \left[b - \beta(b - a), c - \beta(c - b) \right] \tag{7.14}$$

where $y_M(t_0)$ represents the membership grade and $y_N(t_0)$ the non-membership grade in the intuitionistic fuzzy environment.

Considering (α, β)-cut, the modified differential equation (7.5) in an intuitionistic fuzzy environment can be written as:

$$\frac{dy}{dt} = f\left(t_n, \left[\left(\underline{y}(t, \alpha), \overline{y}(t, \alpha) \right), \left(\underline{y}(t, \beta), \overline{y}(t, \beta) \right) \right] \right)$$

with the initial condition $y(t_0)_{\alpha \cdot \beta} = \left[\left(\underline{y}(t_0, \alpha), \overline{y}(t_0, \alpha) \right), \left(\underline{y}(t_0, \beta), \overline{y}(t_0, \beta) \right) \right]$. We use Taylor's theorem to develop Euler's method in an intuitionistic fuzzy environment. Suppose that $y(t)$ is the approximate solution to the given equation in the interval $[t_0, P]$; then using Taylor's theorem, we have

$$y_M(t_{n+1}) = y_M(t_n) + (t_{n+1} - t_n) y_M'(t_n) + \frac{1}{2}(t_{n+1} - t_n)^2 y_M''(\zeta) + \cdots \tag{7.15}$$

$$y_N(t_{n+1}) = y_N(t_n) + (t_{n+1} - t_n) y_N'(t_n) + \frac{1}{2}(t_{n+1} - t_n)^2 y_N''(\zeta) + \cdots \tag{7.16}$$

Here, for each $n = 0, 1, 2, \ldots, N - 1$, equations (7.15) and (7.16) hold well for some numbers $\zeta_n \in (t_n, t_{n+1})$. Taking $h = t_{n+1} - t_n$ in equations (7.15) and (7.16), we get

$$y_M(t_{n+1}) = y_M(t_n) + h y_M'(t_n) + \frac{1}{2} h^2 y_M''(\zeta) + \cdots \tag{7.17}$$

$$y_N(t_{n+1}) = y_N(t_n) + h y_N'(t_n) + \frac{1}{2} h^2 y_N''(\zeta) + \cdots \tag{7.18}$$

Since $y'(t_n) = f(t_n, y(t_n))$ is an approximate solution to the given differential equations (7.17) and (7.18), we can write them as

$$y_M(t_{n+1}) = y_M(t_n) + hf(t_n, y(t_n)) + \frac{1}{2}h^2 y_M''(\zeta) + \cdots \tag{7.19}$$

$$y_N(t_{n+1}) = y_N(t_n) + hf(t_n, y(t_n)) + \frac{1}{2}h^2 y_N''(\zeta) + \cdots \tag{7.20}$$

By truncating from the second-order term in equations (7.19) and (7.20) and denoting $y(t_0) = [\underline{y}(t), \overline{y}(t)]$, we obtain the following equations:

$$\underline{y}_M(t_{n+1}) = \underline{y}_M(t_n) + hf(t_n, \underline{y}(t_n)) \tag{7.21}$$

$$\overline{y}_M(t_{n+1}) = \overline{y}_M(t_n) + hf(t_n, \overline{y}(t_n)) \tag{7.22}$$

$$\underline{y}_N(t_{n+1}) = \underline{y}_N(t_n) + hf(t_n, \underline{y}(t_n)) \tag{7.23}$$

$$\overline{y}_N(t_{n+1}) = \overline{y}_N(t_n) + hf(t_n, \overline{y}(t_n)) \tag{7.24}$$

Here, $y(t_{n+1}) = [\underline{y}_M(t_{n+1}), \overline{y}_M(t_{n+1}); \underline{y}_N(t_{n+1}), \overline{y}_N(t_{n+1})]$ represents the solution in the form of lower and upper bounds of membership at time t_{n+1}. Equations (7.21)–(7.24) represent Euler's method in the fuzzy environment.

7.3.3 Solution to first-order differential equation in fuzzy environment using modified Euler's method

Here, we consider the first-order differential equation (7.5) with fuzzified initial value (7.6), using Taylor's theorem and Euler's method to derive the modified Euler's method in a fuzzy environment.

Suppose that $y(t)$ is the approximate solution to the given equation in the interval $[t_0, P]$; then using Taylor's theorem, we have

$$\underline{y}(t_{n+1}) = \underline{y}(t_n) + (t_{n+1} - t_n)\underline{y}'(t_n) + \frac{1}{2}(t_{n+1} - t_n)^2 \underline{y}''(t_n) + \cdots \tag{7.25}$$

$$\overline{y}(t_{n+1}) = \overline{y}(t_n) + (t_{n+1} - t_n)\overline{y}'(t_n) + \frac{1}{2}(t_{n+1} - t_n)^2 \overline{y}''(t_n) + \cdots \tag{7.26}$$

Here, equations (7.25) and (7.26) are valid for each $n = 0,1,2,\ldots, N-1$, and some numbers $\zeta_n \in (t_n, t_{n+1})$. Taking $h = t_{n+1} - t_n$ in equations (7.25) and (7.26), we get

$$\underline{y}(t_{n+1}) = \underline{y}(t_n) + h\underline{y}'(t_n) + \frac{1}{2}h^2\underline{y}''(t_n) + \cdots \tag{7.27}$$

$$\overline{y}(t_{n+1}) = \overline{y}(t_n) + h\overline{y}'(t_n) + \frac{1}{2}h^2\overline{y}''(t_n) + \cdots \tag{7.28}$$

By truncating the higher-order terms from third order in equations (7.27) and (7.28) and using equation (7.1), we get

$$\underline{y}(t_{n+1}) = \underline{y}(t_n) + h\underline{y}'(t_n) + \frac{1}{2}h^2\left(\frac{\left(\underline{y}'(t_{n+1})\right)\Theta_H\left(\underline{y}'(t_n)\right)}{h}\right) \tag{7.29}$$

$$\overline{y}(t_{n+1}) = \overline{y}(t_n) + h\overline{y}'(t_n) + \frac{1}{2}h^2\left(\frac{\left(\overline{y}'(t_{n+1})\right)\Theta_H\left(\overline{y}'(t_n)\right)}{h}\right) \tag{7.30}$$

In equations (7.29) and (7.30), $y'(t)$ is replaced using equation (7.5), i.e., $y'(t) = f(t_n, y(t_n))$

$$\underline{y}(t_{n+1}) = \underline{y}(t_n) + \frac{h}{2}\left(f(t_{n+1}, \underline{y}(t_{n+1}))\Theta_H f(t_n, \underline{y}(t_n))\right) \tag{7.31}$$

$$\overline{y}(t_{n+1}) = \overline{y}(t_n) + \frac{h}{2}\left(f(t_{n+1}, \overline{y}(t_{n+1}))\Theta_H f(t_n, \overline{y}(t_n))\right) \tag{7.32}$$

In order to obtain the solution using numerical approach, it is necessary to know the value of $(y(t_{n+1}))$ appearing on the right-hand side. To find out the value of $(y(t_{n+1}))$, we first predict it by Euler's method and then use modified Euler's method to correct it. Here, the modified Euler's method for the fuzzy environment is written as:

$$\begin{cases} \underline{y}^*(t_{n+1}) = \underline{y}(t_n) + h\,f(t_n, \underline{y}(t_n)) \\ \underline{y}(t_{n+1}) = \underline{y}(t_n) + \frac{h}{2}\left(f(t_{n+1}, \underline{y}^*(t_{n+1})) + f(t_n, \underline{y}(t_n))\right) \end{cases} \tag{7.33}$$

$$\begin{cases} \overline{y}^*(t_{n+1}) = \overline{y}(t_n) + h\, f\left(t_n, \overline{y}(t_n)\right) \\ \overline{y}(t_{n+1}) = \overline{y}(t_n) + \dfrac{h}{2}\left(f\left(t_{n+1}, \overline{y}^*(t_{n+1})\right) + f\left(t_n, \overline{y}(t_n)\right)\right) \end{cases} \qquad (7.34)$$

where $n = 0,1, 2 \ldots N{-}1$, and $y(t_{n+1}) = \left[\underline{y}(t_{n+1}), \overline{y}(t_{n+1})\right]$ represents the solution in the form of fuzzy environment.

7.3.4 Solution to first-order differential equation in intuitionistic fuzzy environment using modified Euler's method

Here, we consider the first-order differential equation (7.12) with fuzzified initial value (13–14), using Taylor's theorem and Euler's method to derive the modified Euler's method in a fuzzy environment.

Suppose that $y(t)$ is the approximate solution to the given equation in the interval $[t_0, P]$; then using Taylor's theorem, we have

$$\underline{y_M}(t_{n+1}) = \underline{y_M}(t_n) + (t_{n+1} - t_n)\underline{y_M'}(t_n) + \frac{1}{2}(t_{n+1} - t_n)^2\, \underline{y_M''}(t_n) + \cdots \qquad (7.35)$$

$$\overline{y_M}(t_{n+1}) = \overline{y_M}(t_n) + (t_{n+1} - t_n)\overline{y_M'}(t_n) + \frac{1}{2}(t_{n+1} - t_n)^2\, \overline{y_M''}(t_n) + \cdots \qquad (7.36)$$

$$\underline{y_N}(t_{n+1}) = \underline{y_N}(t_n) + (t_{n+1} - t_n)\underline{y_N'}(t_n) + \frac{1}{2}(t_{n+1} - t_n)^2\, \underline{y_N''}(t_n) + \cdots \qquad (7.37)$$

$$\overline{y_N}(t_{n+1}) = \overline{y_N}(t_n) + (t_{n+1} - t_n)\overline{y_N'}(t_n) + \frac{1}{2}(t_{n+1} - t_n)^2\, \overline{y_N''}(t_n) + \cdots \qquad (7.38)$$

Here, equations (7.22)–(7.25) are valid for each $n = 0,1,2,\ldots, N-1$, and some numbers $\zeta_n \in (t_n, t_{n+1})$. Taking $h = t_{n+1} - t_n$ in equations (7.22)–(7.25), we get

$$\underline{y_M}(t_{n+1}) = \underline{y_M}(t_n) + h\underline{y_M'}(t_n) + \frac{1}{2}h^2\, \underline{y_M''}(t_n) + \cdots \qquad (7.39)$$

$$\overline{y_N}(t_{n+1}) = \overline{y_N}(t_n) + h\overline{y_N'}(t_n) + \frac{1}{2}h^2\, \overline{y_N''}(t_n) + \cdots \qquad (7.40)$$

$$\underline{y_N}\left(t_{n+1}\right)=\underline{y_N}\left(t_n\right)+h\underline{y_N'}\left(t_n\right)+\frac{1}{2}h^2\,\underline{y_N''}\left(t_n\right)+\cdots \tag{7.41}$$

$$\overline{y_M}\left(t_{n+1}\right)=\overline{y_M}\left(t_n\right)+h\overline{y_M'}\left(t_n\right)+\frac{1}{2}h^2\,\overline{y_M''}\left(t_n\right)+\cdots \tag{7.42}$$

By truncating from third-order term in equations (7.39)–(7.42) and using equation (7.1), we get

$$\underline{y_M}\left(t_{n+1}\right)=\underline{y_M}\left(t_n\right)+h\underline{y_M'}\left(t_n\right)+\frac{1}{2}h^2\left(\underline{y_M'}\left(t_{n+1}\right)\right)\theta_H\left(\underline{y_M'}\left(t_n\right)\right) \tag{7.43}$$

$$\overline{y_M}\left(t_{n+1}\right)=\overline{y_M}\left(t_n\right)+h\overline{y_M'}\left(t_n\right)+\frac{1}{2}h^2\left(\overline{y_M'}\left(t_{n+1}\right)\right)\theta_H\left(\overline{y_M'}\left(t_n\right)\right)\mathrm{P} \tag{7.44}$$

$$\underline{y_N}\left(t_{n+1}\right)=\underline{y_N}\left(t_n\right)+h\underline{y_N'}\left(t_n\right)+\frac{1}{2}h^2\left(\underline{y_N'}\left(t_{n+1}\right)\right)\theta_H\left(\underline{y_N'}\left(t_n\right)\right) \tag{7.45}$$

$$\overline{y_N}\left(t_{n+1}\right)=\overline{y_N}\left(t_n\right)+h\overline{y_N'}\left(t_n\right)+\frac{1}{2}h^2\left(\overline{y_N'}\left(t_{n+1}\right)\right)\theta_H\left(\overline{y_N'}\left(t_n\right)\right) \tag{7.46}$$

In equations (7.26) and (7.27), $y'(t)$ is replaced using equation (7.5), i.e., $y'(t)=f\left(t_n,y(t_n)\right)$

$$\underline{y_M}\left(t_{n+1}\right)=\underline{y_M}\left(t_n\right)+\frac{h}{2}\,f\left(t_{n+1},\underline{y_M}\left(t_{n+1}\right)\right)\theta_H\,f\left(t_n,\underline{y_M}\left(t_n\right)\right) \tag{7.47}$$

$$\overline{y_M}\left(t_{n+1}\right)=\overline{y}_M\left(t_n\right)+\frac{h}{2}\,f\left(t_{n+1},\overline{y_M}\left(t_{n+1}\right)\right)\theta_H\,f\left(t_n,\overline{y_M}\left(t_n\right)\right) \tag{7.48}$$

$$\underline{y_N}\left(t_{n+1}\right)=\underline{y_N}\left(t_n\right)+\frac{h}{2}\,f\left(t_{n+1},\underline{y_N}\left(t_{n+1}\right)\right)\theta_H\,f\left(t_n,\underline{y_N}\left(t_n\right)\right) \tag{7.49}$$

$$\overline{y}_N\left(t_{n+1}\right)=\overline{y}_N\left(t_n\right)+\frac{h}{2}\,f\left(t_{n+1},\overline{y_N}\left(t_{n+1}\right)\right)\theta_H\,f\left(t_n,\overline{y_N}\left(t_n\right)\right) \tag{7.50}$$

In order to obtain the solution using numerical approach, it is necessary to know the value of $\left(y(t_{n+1})\right)_{M,N}$ appearing on the right-hand side. To find out the value of $\left(y(t_{n+1})\right)_{M,N}$, we first predict it by Euler's method and then use modified Euler's method to correct it. Here, the modified Euler's method for the fuzzy environment is written as:

$$\begin{cases} \overset{*}{y}_M(t_{n+1}) = \underline{y}_M(t_n) + h\, f\big(t_n, \underline{y}_M(t_n)\big) \\ \underline{y}_M(t_{n+1}) = \underline{y}_M(t_n) + \dfrac{h}{2}\Big(f\big(t_{n+1}, \overset{*}{y}_M(t_{n+1})\big) + f\big(t_n, \underline{y}_M(t_n)\big)\Big) \end{cases} \tag{7.51}$$

$$\begin{cases} \overline{\overset{*}{y}}_M(t_{n+1}) = \overline{y}_M(t_n) + h\, f\big(t_n, \overline{y}_M(t_n)\big) \\ \overline{y}_M(t_{n+1}) = \overline{y}_M(t_n) + \dfrac{h}{2}\Big(f\big(t_{n+1}, \overline{\overset{*}{y}}_M(t_{n+1})\big) + f\big(t_n, \overline{y}_M(t_n)\big)\Big) \end{cases} \tag{7.52}$$

$$\begin{cases} \overset{*}{y}_N(t_{n+1}) = \underline{y}_N(t_n) + h\, f\big(t_n, \underline{y}_N(t_n)\big) \\ \underline{y}_N(t_{n+1}) = \underline{y}_N(t_n) + \dfrac{h}{2}\Big(f\big(t_{n+1}, \overset{*}{y}_N(t_{n+1})\big) + f\big(t_n, \underline{y}_N(t_n)\big)\Big) \end{cases} \tag{7.53}$$

$$\begin{cases} \overline{\overset{*}{y}}_N(t_{n+1}) = \overline{y}_N(t_n) + h\, f\big(t_n, \overline{y}_N(t_n)\big) \\ \overline{y}_N(t_{n+1}) = \overline{y}_N(t_n) + \dfrac{h}{2}\Big(f\big(t_{n+1}, \overline{\overset{*}{y}}_N(t_{n+1})\big) + f\big(t_n, \overline{y}_N(t_n)\big)\Big) \end{cases} \tag{7.54}$$

where $n = 0, 1, 2 \ldots N-1$, and $y(t_{n+1})_{M,N} = \left[\underline{y}(t_{n+1})_{M,N}, \overline{y}(t_{n+1})_{M,N} \right]$ represents the solution in the form of membership and non-membership at time t_{n+1}.

7.4 ILLUSTRATION

Let us consider a differential equation as

$$y'(t) = y(t) + \overset{\cdot}{1} \tag{7.55}$$

where y is the dependent variable, t is the independent variable, and $\overset{\cdot}{1}$ is the constant coefficient in the form of fuzzy and intuitionistic fuzzy numbers with initial values $y(0) = 1$. We need to calculate the value of $y(t)$ at $t = 1$.

Solution: We apply the classical, Euler's, and modified Euler's methods in fuzzy and intuitionistic fuzzy environments and then compare the solutions.

Method I: Solution Obtained by Classical Method

Solving first-order linear differential equation (7.55) with initial condition $y(0) = 1$, we get the following solution:

$$y(t) = 2e^t - 1 \tag{7.56}$$

By substituting $t = 1$ in equation (7.56), the solution to $y(t)$ is 4.4366 up to four decimal places.

Here, the same problem is solved using Euler's method and modified Euler's method by taking initial condition and constant parameter 1 in fuzzy and intuitionistic fuzzy environments.

Method 2: Solution Obtained by Euler's Method

Case 1: Using Initial Condition and Constant Parameter in Fuzzy Triangular Form

Let us consider uncertainty in constant parameter $\tilde{1}$, and hence, we get

$$\tilde{1} = [0.8 + 0.2a, 1.2 - 0.2a] \tag{7.57}$$

The differential equation in the form fuzzy triangular number is given below:

$$y'(t) = y(t) + [0.8 + 0.2a, 1.2 - 0.2a] \tag{7.58}$$

where a represents the α-cut for constant parameter 1 and the initial value is

$$y(0) = [0.6 + 0.4\alpha, 1.4 - 0.4\alpha] \tag{7.59}$$

Solving first-order differential equations in the form of triangular fuzzy numbers using the proposed numerical technique, i.e., Euler's method in the fuzzy environment, we get the following equations:

$$\underline{y}(t_{n+1}) = \underline{y}(t_n) + (0.1)\left(\underline{y}(t_n) + [0.8 + 0.2a]\right) \tag{7.60}$$

$$\overline{y}(t_{n+1}) = \overline{y}(t_n) + (0.1)\left(\overline{y}(t_n) + [1.2 - 0.2a]\right) \tag{7.61}$$

where step size $h = 0.1$, $n = 0, 1, 2...9$, and higher-order terms are ignored.

Equations (7.60) and (7.61) are solved using fuzzy triangular form for constant parameter, i.e., $\tilde{1} = [0.8 + 0.2a, 1.2 - 0.2a]$, and initial values, i.e., $y(0) = [0.6 + 0.4\alpha, 1.4 - 0.4\alpha]$. For $n = 0, 1...9$, we obtained values of $y(t)$ for different α-cut values with step size 0.1. In Table 7.1, we calculated numerical values for $y(t)$ for different α-cut values at $t = 1$.

From Table 7.1, the value of $y(t)$ for lower bound increases and the upper bound decreases. For α-cut and $a-$ cut, value of $y(t)$ for $a = 1$ and $\alpha = 1$ the fuzzy solution matches with crisp solution.

Table 7.1 Fuzzy solution to y(t) using Euler's method at $t = 1$ with respect to α-cut

α – cut	a – cut	Lower bound of truth value at $t = 1$ $\underline{y}(t_0)$	Upper bound of truth value at $t = 1$ $\overline{y}(t_0)$
0	0	2.831239	5.54373
0.2	0	3.038739	5.336231
0.4	0	3.246238	5.128732
0.6	0	3.453738	4.921232
0.8	0	3.661237	4.713733
1	0	3.868736	4.506233
0	1	3.149988	5.224982
0.2	1	3.357487	5.017483
0.4	1	3.564987	4.809983
0.6	1	3.772486	4.602484
0.8	1	3.979986	4.394984
1	1	4.187485	4.187485

Case 2 Using Initial Condition and Constant Parameter in an Intuitionistic Fuzzy Triangular Form

Let us consider uncertainty in constant parameter $\breve{1}$, and hence, we get

$$\breve{1} = [0.8 + 0.2a, 1.2 - 0.2a]_M, \ [1 - 0.1b, 1 + 0.1b]_N \tag{7.62}$$

The differential equation in the form fuzzy triangular number is given below:

$$y'(t)_M = y(t) + [0.8 + 0.2a, 1.2 - 0.2a]_M$$
$$y'(t)_N = y(t) + [1 - 0.1b, 1 + 0.1b]_N \tag{7.63}$$

where a represents (α, β)-cut for constant parameter $\breve{1}$, respectively, and initial values are

$$y(0)_M = [0.6 + 0.4\alpha, 1.4 - 0.4\alpha], \ y(0)_N = [1 - 0.3\beta, 1 + 0.3\beta] \tag{7.64}$$

Solving first-order differential equations in the form of triangular intuitionistic fuzzy numbers using the proposed numerical technique, i.e., Euler's method in intuitionistic fuzzy environment, we get the following equations:

$$\underline{y}(t_{n+1})_M = \underline{y}(t_n)_M + (0.1)\left(\underline{y}(t_n)_M + [0.8 + 0.2a]_M \right) \tag{7.65}$$

$$\overline{y}(t_{n+1})_M = \overline{y}(t_n)_M + (0.1)\left(\overline{y}(t_n)_M + [1.2 - 0.2a]_M\right) \qquad (7.66)$$

$$\underline{y}(t_{n+1})_N = \underline{y}(t_n)_N + (0.1)\left(\underline{y}(t_n)_N + [1 - 0.1b]_N\right) \qquad (7.67)$$

$$\overline{y}(t_{n+1})_N = \overline{y}(t_n)_N + (0.1)\left(\overline{y}(t_n)_N + [1 + 0.1b]_N\right) \qquad (7.68)$$

where step size $h = 0.1$, $n = 0,1,\ 2...9$, and the higher-order terms are ignored.

Equations (7.65)–(7.68) are solved using initial values and constant parameter in the form of intuitionistic fuzzy triangular number. We obtained the values of $y(t)$ for different (α, β)-cut values with step size 0.1. In Table 7.2, we calculated numerical values for $y(t)$ for different (α, β)-cut values at $t = 1$.

From Table 7.2, the value of membership $y(t)_M$ for lower bound increases and that for the upper bound decreases. Similarly, the value of non-membership $y(t)_N$ for lower bound decreases and that for upper bound increases. For (α, β)-cut and (a,b)– cut, value of membership $y(t)_T$ for $a = 1$ and $\alpha = 1$, the intuitionistic solution matches with crisp solution. Similarly, the value of non-membership $y(t)_N$ matches with crisp solution at $b = 0$ and $\beta = 0$.

Table 7.2 Intuitionistic fuzzy solution to y(t) using Euler's method at t = 1 with respect to (α, β)—cut

(α, β)-cut	(a,b) – cut	Lower bound value of $\underline{y}(t_0)_M$	Upper bound value of $\overline{y}(t_0)_M$	Lower bound value of $\underline{y}(t_0)_N$	Upper bound value of $\overline{y}(t_0)_N$
0	0	2.831239	5.543730	4.187485	4.187485
0.2	0	3.038739	5.336231	4.031866	4.343109
0.4	0	3.246238	5.128732	3.876236	4.498734
0.6	0	3.453738	4.921232	3.720611	4.654359
0.8	0	3.661237	4.713733	3.564987	4.809983
1	0	3.868736	4.506233	3.409362	4.965608
0	1	3.149988	5.224982	4.02811112	4.34685911
0.2	1	3.357487	5.017483	3.87248621	4.50248412
0.4	1	3.564987	4.809983	3.71686222	4.65810821
0.6	1	3.772486	4.602484	3.56123724	4.81373322
0.8	1	3.979986	4.394984	3.40561225	4.96935723
1	1	4.187485	4.187485	3.24998812	5.12498256

Method 3: Solution Obtained by Modified Euler's Method

Case I: Using Initial Condition and Constant Parameter in Fuzzy Triangular Form

Here, we consider uncertainty in constant parameter $\tilde{1}$ (see 7.57), and hence, we get

$$\tilde{1} = [0.8 + 0.2a, 1.2 - 0.2a]$$

The differential equation in the form of fuzzy triangular number is given below:

$$y'(t) = y(t) + [0.8 + 0.2a, 1.2 - 0.2a]$$

where a represents α-cut for constant parameter $\tilde{1}$, respectively, and the initial value (see 7.59) is

$$y(0) = [0.6 + 0.4\alpha, 1.4 - 0.4\alpha]$$

Hence, solving the first-order differential equation using the numerical technique, i.e., modified Euler's method in the fuzzy environment, we get the following solution:

$$\begin{cases} \underline{y}^*(t_{n+1}) = \underline{y}(t_n) + (0.1)\left(\underline{y}(t_n) + [0.8 + 0.2a]\right) \\ \underline{y}(t_{n+1}) = \underline{y}(t_n) + 0.05\left(\underline{y}^*(t_{n+1}) + \left(\underline{y}(t_n) + [0.8 + 0.2a]\right)\right) \end{cases} \tag{7.69}$$

$$\begin{cases} \overline{y}^*(t_{n+1}) = \overline{y}(t_n) + (0.1)\left(\overline{y}(t_n) + [1.2 - 0.2a]\right) \\ \overline{y}(t_{n+1}) = \overline{y}(t_n) + 0.05\left(\overline{y}^*(t_{n+1}) + \left(\overline{y}(t_n) + [1.2 - 0.2a]\right)\right) \end{cases} \tag{7.70}$$

where step size $h = 0.1$ and $n = 0, 1, 2 \ldots 9$.

Equations (7.69) and (7.70) are solved using the initial value and constant parameter in the form of fuzzy triangular number. For $n = 0, 1 \ldots 9$, we obtained predicted values of lower and upper bound membership using Euler's method at different α-cut values with step size $h = 0.1$ and then we used modified Euler's method to obtain corrector values for the same. In Table 7.3, we calculated numerical values $y(t)$ at different α-cut values with step size $h = 0.1$ at $t = 1$.

From Table 7.3, the value of membership $y(t)$ for lower bound increases and the upper bound decreases. In addition, from Table 7.1, we observe that the value of lower and upper bounds of truth membership is 4.187485 for

α-cut and $a-$cut equal to 1, and on the other hand, in Table 7.3, the value of lower and upper bounds of truth membership is 4.428162 for (α, β, γ)-cut and $(a,b,c)-$cut equal to 1.

Case 2 Using Initial Condition and Constant Parameter in an Intuitionistic Fuzzy Triangular Form

Here, we consider uncertainty in constant parameter $\tilde{1}$ (see 7.62), and hence, we get

$$\tilde{1} = [0.8+0.2a,1.2-0.2a]_M ,[1-0.1b,1+0.1b]_N$$

The differential equation in the form of fuzzy triangular number is given below:

$$y'(t)_M = y(t)+ [0.8+0.2a,1.2-0.2a]_M$$

$$y'(t)_N = y(t)+ [1-0.1b,1+0.1b]_N$$

where a represents (α, β)-cut for constant parameter 1, respectively, and initial values (see 7.64) are

$$y(0)_M = [0.6+0.4\alpha,1.4-0.4\alpha], \ y(0)_N = [1-0.3\beta,1+0.3\beta]$$

Hence, solving the first-order differential equation using the numerical technique, i.e., modified Euler's method in the fuzzy environment, we get the following solution:

Table 7.3 Fuzzy solution to y(t) using Euler's method at $t = 1$ with respect to α–cut

α - cut	α - cut	Lower bound of truth value at $t = 1$ $\underline{y}(t_0)$	Upper bound of truth value at $t = 1$ $\overline{y}(t_0)$
0	0	2.999713	5.551505
0.2	0	3.216840	5.334379
0.4	0	3.433966	5.117253
0.6	0	3.651093	4.900126
0.8	0	3.868219	4.683021
1	0	4.085346	4.465873
0	1	3.342529	5.513794
0.2	1	3.559656	5.296668
0.4	1	3.776782	5.079541
0.6	1	3.993909	4.862415
0.8	1	4.211035	4.645288
1	1	4.428162	4.428162

$$\begin{cases} \underline{y}^{*}\left(t_{n+1}\right)_{M} = \underline{y}\left(t_{n}\right)_{M} + (0.1)\left(\underline{y}\left(t_{n}\right)_{M} + \left[0.8 + 0.2a\right]_{M}\right) \\ \underline{y}\left(t_{n+1}\right)_{M} = \underline{y}\left(t_{n}\right)_{M} + 0.05\left(\underline{y}^{*}\left(t_{n+1}\right)_{M} + \left(\underline{y}\left(t_{n}\right)_{M} + \left[0.8 + 0.2a\right]_{M}\right)\right) \end{cases} \tag{7.71}$$

$$\begin{cases} \overline{y}^{*}\left(t_{n+1}\right)_{M} = \overline{y}\left(t_{n}\right)_{M} + (0.1)\left(\overline{y}\left(t_{n}\right)_{M} + \left[1.2 - 0.2a\right]_{M}\right) \\ \overline{y}\left(t_{n+1}\right)_{M} = \overline{y}\left(t_{n}\right)_{M} + 0.05\left(\overline{y}^{*}\left(t_{n+1}\right)_{M} + \left(\overline{y}\left(t_{n}\right)_{M} + \left[1.2 - 0.2a\right]_{M}\right)\right) \end{cases} \tag{7.72}$$

$$\begin{cases} \underline{y}^{*}\left(t_{n+1}\right)_{N} = \underline{y}\left(t_{n}\right)_{N} + (0.1)\left(\underline{y}\left(t_{n}\right)_{N} + \left[1 - 0.1b\right]_{N}\right) \\ \underline{y}\left(t_{n+1}\right)_{N} = \underline{y}\left(t_{n}\right)_{N} + 0.05\left(\underline{y}^{*}\left(t_{n+1}\right)_{N} + \left(\underline{y}\left(t_{n}\right)_{N} + \left[1 - 0.1b\right]_{N}\right)\right) \end{cases} \tag{7.73}$$

$$\begin{cases} \overline{y}^{*}\left(t_{n+1}\right)_{N} = \overline{y}\left(t_{n}\right)_{N} + (0.1)\left(\overline{y}\left(t_{n}\right)_{N} + \left[1 + 0.1b\right]_{N}\right) \\ \overline{y}\left(t_{n+1}\right)_{N} = \overline{y}\left(t_{n}\right)_{N} + 0.05\left(\overline{y}^{*}\left(t_{n+1}\right)_{N} + \left(\overline{y}\left(t_{n}\right)_{N} + \left[1 + 0.1b\right]_{N}\right)\right) \end{cases} \tag{7.74}$$

where step size $h = 0.1$ and $n = 0,1, 2...9$.

Equations (7.71)–(7.74) are solved using the initial value and constant parameter in the form of intuitionistic fuzzy triangular number. For $n = 0$, $1...9$, we obtained predicted values of lower and upper bound membership using Euler's method at different (α, β)-cut values with step size $h = 0.1$ and then we used modified Euler's method to obtain corrector values for the same. In Table 7.4, we calculated numerical values $y(t)$ at different (α, β)-cut values with step size $h = 0.1$ at $t = 1$.

From Table 7.4, the value of membership $y(t)_{M}$ for the lower bound increases and that for the upper bound decreases. Similarly, the value of non-membership $y(t)_{N}$ for the lower bound decreases and that for the upper bound increases. For (α, β)-cut and $(a,b)-$ cut, value of membership $y(t)_{T}$ for $a = 1$ and $\alpha = 1$ the intuitionistic solution matches with crisp solution. Similarly, the value of non-membership $y(t)_{N}$ matches with crisp solution at $b = 0$ and $\beta = 0$.

7.5 CONCLUSIONS

In this chapter, we developed theories of Euler's method and modified Euler's method in a fuzzy environment as well as intuitionistic environment to obtain the solution to first-order differential equation. For that, we considered the first-order ordinary differential equation with initial conditions and constant parameter in the form of triangular number in fuzzy

Table 7.4 Intuitionistic fuzzy solution to $y(t)$ using modified Euler's method at $t = 1$ with respect to (α, β)-cut

(α, β)-cut	(a,b)-cut	Lower bound value of $\underline{y}(t_0)_M$	Upper bound value of $\overline{y}(t_0)_M$	Lower bound value of $\underline{y}(t_0)_N$	Upper bound value of $\overline{y}(t_0)_N$
0	0	2.999713	5.551505	4.428162	4.428162
0.2	0	3.21684	5.334379	4.265317	4.591007
0.4	0	3.433966	5.117253	4.102472	4.753851
0.6	0	3.651093	4.900126	3.939627	4.916696
0.8	0	3.868219	4.683	3.776782	5.079541
1	0	4.085346	4.465873	3.613937	5.242386
0	1	3.342529	5.513794	4.256754	4.59957
0.2	1	3.559656	5.296668	4.093909	4.762415
0.4	1	3.776782	5.079541	3.931064	4.925259
0.6	1	3.993909	4.862415	3.768219	5.088104
0.8	1	4.211035	4.645288	3.605374	5.250949
1	1	4.428162	4.428162	3.442529	5.413794

and intuitionistic environments. To show the effectiveness of the proposed method, it has been applied to general examples where the solution is given in terms of the membership grade in the fuzzy environment and both membership and non-membership grade in the intuitionistic environment. We have shown the results in tables for different α-cut values for fuzzy solution and (α, β)-cut. The results obtained are also discussed in detail.

REFERENCES

1. Zadeh, L. A.; Fuzzy sets, Information and Control, 1965, Volume 8, No. 3, pp. 338–353.
2. Atanassov, K.T.; Intuitionistic fuzzy sets, *Fuzzy Sets and Systems*, 1986, Volume 20, No. 1, pp. 87–96.
3. Dubois, D., Prade, H.; Towards fuzzy differential calculus part 2: Integration on fuzzy intervals, *Fuzzy Sets and Systems*, 1982, Volume 8, 105–115.
4. Dubois, D., Prade, H.; Towards fuzzy differential calculus part 3: Differentiation, *Fuzzy Sets and Systems*, 1982, Volume 8, No. 3, pp. 225–233.
5. Dubois, D., Prade, H.; Operations on fuzzy numbers, *International Journal of Systems Science*, 1978, Volume 9, No. 6, pp. 613–626.
6. Chang, S.S.L., Zadeh, L.A.; On fuzzy mapping and control, *IEEE Transactions on Systems, Man and Cybernetics*, 1972, Volume 2, No. 1, pp. 30–34.
7. B. Ben Amma and L. S. Chadli; Numerical solution of intuitionistic fuzzy differential equations by Runge-Kutta Method of order four, *Notes on Intuitionistic Fuzzy Sets*, 2016, Volume 22, No.4, pp. 42–52.

8. B. Ben Amma, S. Melliani and L. S. Chadli; *Intuitionistic Fuzzy Functional Differential Equations, Fuzzy Logic in Intelligent System Design: Theory and Applications*, Ed. Cham: Springer International Publishing, 2018, pp.335–357.

9. Puri, M. L., Ralescu, D. A.; Differentials for fuzzy functions, *J. Mathematical Anal. & Appl.*, 1983, Volume 91, No. 2, pp. 552–558.

10. Goetschel, R., Voxman, W.; Elementary fuzzy calculus, *Fuzzy Sets and Systems*, 1986, Volume 18, No. 1, pp. 31–43.

11. Seikkala, S.; On the fuzzy initial value problem, *Fuzzy Sets and Systems*, 1987, Volume 24, No. 3, pp. 319–330.

12. Kaleva, O.; Fuzzy differential equations, *Fuzzy Sets and Systems*, 1987, Volume 24, pp. 301–317.

13. Buckley, J. J., Feuring, T.; Fuzzy differential equations, *Fuzzy Sets and Systems*, 2000, Volume 110, pp. 43–54.

14. Buckley J. J., Feuring, T.; Fuzzy initial value problem for nth-order linear differential equations, *Fuzzy Sets and Systems*, 2001, Volume 121, pp. 247–255.

15. Bede B., Gal, S. G.; Generalizations of the differentiability of fuzzy-number valued functions with applications to fuzzy differential equations, *Fuzzy Sets and Systems*, 2005, Volume 151, No. 3, pp. 581–599.

16. L. Stefanini and B. Bede, Generalized Hukuhara differentiability of interval-valued functions and interval differential equations, *Nonlinear Analysis, Theory, Methods & Applications*, 2009, Volume 71, No. 3–4, pp. 1311–1328.

17. Salahshour S., Allahviranloo T., Applications of fuzzy Laplace transforms, *Soft Computing*, 2013, Volume 17, No. 1, pp. 145–158.

18. Mondal, S. P., Banerjee, S., Roy, T. K. First order linear homogeneous ordinary differential equation in fuzzy environment, *International Journal of Pure and Applied Sciences*, 2013, Volume14, No. 1, pp. 16–26.

19. Mondal S. P., Roy, T. K. First order linear homogeneous ordinary differential equation in fuzzy environment based on Laplace transform, *Journal of Fuzzy Set Valued Analysis*, 2013, Volume 2013, pp. 1–18.

20. Ahmadian A., Salahshour S., Chan C. S., and Baleanu D., Numerical solutions of fuzzy differential equations by an efficient Runge–Kutta method with generalized differentiability, *Fuzzy Sets and Systems*, 2018, Volume 331, pp. 47–67.

21. Tapaswini, S.; Chakraverty, S.; Allahviranloo, T.; A new approach to nth order fuzzy differential equations, *Computational Mathematics and Modeling*, 2017, Volume 28, No. 2, pp. 278–300.

22. Sahni M., Sahni R., Verma R., Mandaliya A., Shah D.; Second order Cauchy Euler equation and its application for finding radial displacement of a solid disk using generalized trapezoidal intuitionistic fuzzy number, *WSEAS Transactions on Mathematics*, 2019, Volume 18, pp. 37–45.

23. Sahni, M., Parikh, M., Sahni, R.; Sumudu transform for solving ordinary differential equation in a fuzzy environment. *Journal of Interdisciplinary Mathematics*, 2021, Volume 24, pp. 1–13.

24. Parikh, M., Sahni, M., Sahni, R.; Modelling of mechanical vibrating system in classical and fuzzy Environment using sumudu transform method, *Structural Integrity and Life*, 2020, Volume 20, pp. S54–S60.

Part II

Simulations in machine learning and image processing

Chapter 8

Multi-layer encryption algorithm for data integrity in cloud computing

Arwa Zabian, Shaker Mrayyen, Abram Magdy Jonan, Tareq Al-shaikh, and Mohammed Ghazi Al-Khaiyat

Jadara University

CONTENTS

8.1 INTRODUCTION

The introduction of cloud computing in all life aspects has changed the way of handling each thing in our life. Today, with cloud computing, the possibility of losing or damaging our important data such as names, photos, and important documents is decreased, which adds many benefits on data handling such as flexibility, availability, and capacity. Data are becoming available anytime, anywhere from any device at any location, which means cloud computing has eliminated the time and location and space constraints for everything, including business, education, entertainment, and medicine, that increase the user satisfaction. Cloud applications are increased day after day, all these applications has a single goal that is to provide more facilities to the user. Also, data documenting is becoming more efficient. In a local device, the user has a complete control on his/her data; however, when data are stored on a device on the cloud, he/she can access his/her data at any time, from anywhere, he/she can modify them, retrieve them, stored them, and delete them at any time, but he/she doesn't have complete control on it. To enjoy all the services offered by the cloud computing, it is necessary to apply severe conditions that maintain data integrity, privacy, and availability. Data security is becoming a persistent need in the cloud to give trust in it and to encourage people to use the cloud as a secure way of communication or storage. Undoubtedly, the introduction of cloud has made all aspects of our life easy to handle, such as business, shopping, education, and research, in addition to adding a huge quantity of information. To benefit from these facilities, information security is an essential requirement.

As more people work on information security, so more people try to access the data in an illegal manner and try in different manners to breach the system's security. Encryption is an easy way that helps in protecting our data to be accessed only by authorized persons. Cloud computing is composed of hardware, software, storage, network, interface, and services that provide the user facilities. Applications and services are on demand and are independent of user location. Cloud computing reduces the hardware cost, increases the storage space, and gives maximum capability with minimum resources, which means providing services to the user with the best use of resources. To allow the user to utilize all these benefits of cloud computing, his/her data must be preserved from change, modification, intercept, and any other operation that can breach the data security. When we speak about cloud, we speak about three-tier architecture, in which the client part is the first tier, the server part is the second tier, and the database part is the third tier; data security is important in each part (Figure 8.1). Data privacy is an aspect of information technology that deals with the ability to determine who can read and access our data while they are in transmission over the network or stored on the cloud server.

Encryption is considered one of the tools used for data protection, encryption means converting the message called plain text into incomprehensive

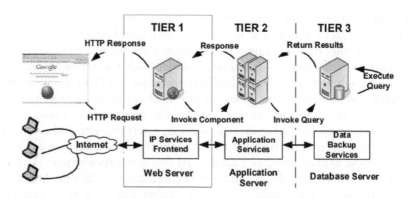

Figure 8.1 Typical three-tier architecture.

text using some encryption algorithms, the goal of encryption algorithms is to make the plain text unclear for unauthorized persons, and the authorized persons can use a decryption algorithm to decipher the text and read it. The decryption process is the reverse process of encryption. There are two types of encryption: symmetric encryption in which only one key is used for encryption and decryption and asymmetric encryption in which two keys are used, where one key is used for encryption and the other different key is used for decryption. In this paper, we propose a multilevel symmetric encryption/decryption algorithm for maintaining data integrity in cloud computing. As mentioned previously, in cloud, data are subject to many threats while being transmitted over the network or stored on the server. Data integrity means data can be accessed and modified only by the authorized persons.

The encryption algorithm that can be one of the following: RSA, DES, AES, Blowfish [1]. Once the encryption algorithm is chosen, it cannot be changed, but the user can view, modify, delete, and edit the file when it is uploaded before encryption. This chapter does not present any simulation results or any analysis of the system performance. Considering the open challenges in cryptographic operations in the cloud, some basic measures must be taken into consideration to protect data against cyber-attacks, such as preventing direct access to data and maintaining data encryption. There are some software packages that prevent attacks called WannaKiwi software [2], which is based on Wana encrypt method that relies on scanning the address space of the process that generates private user key and can recover the key and the files in case of attacks. It is not perfect, but is one of the best solutions in case of victims without backup data. In [3], a public key algorithm ECC (elliptic curve cryptography) is proposed, which is an open key encryption algorithm that has many advantage over RSA. In this chapter, we are interested in presenting a three-tier encryption/decryption algorithm that allows the user data to be transmitted encrypted and stored encrypted on the cloud server. In our proposed system, data are transferred, stored, and retrieved encrypted and only the authorized who have the decryption keys can read and modify the data. Encrypting data more than one time with different methods and different keys will increase the number of attempts needed to crack the key and will make the access to the plain text more difficult.

Problem formulation: Many encryption algorithms are proposed to maintain data integrity and privacy, but all these algorithms work separately and concern on working between sender and receiver, where data are encrypted in one part and decrypted in the other part. In our work, we will use three of these algorithms consecutively to make accessing the data more difficult and complicated. Penetrating the data in our system requires knowing many information; it requires cracking three keys and knowing how three encryption algorithms will work.

The rest of the chapter is organized as follows: Section 8.2 presents related works. Section 8.3 presents our proposed algorithm and the algorithm analysis. Section 8.4 presents simulation results. And finally, Section 8.5 provides conclusions and future work.

8.2 RELATED WORKS

The mobile applications increase day after day, and the use of cloud application is becoming necessary today. So, the users today are involved in all cloud computing layers where his/her sensitive data such as pictures, personal information, and banking information are stored; for that, data security is becoming a necessity and a key requirement for all of us today. In [4], a cipher cloud algorithm is proposed, which allows the user to upload his/her file to cipher cloud to encrypt it and then transfer via an encrypted HTTPS connection to the server. Then, it is decrypted on the server and stored in its original form. The file is uploaded encrypted, but retrieved in its original form; in addition the user can choose the homomorphic encryption, the proposed solution can be used to maintain data security and confidentiality, where mathematical operations are performed on encrypted data without using original data. Data are encrypted and decrypted in client side, and the service provider works on cipher text only. In [5], it is shown that the homomorphic encryption can provide the same level of security of any encryption system and can maintain a complete privacy between client and cloud server without decreasing system functionality. In [6], two times encryption method is proposed, which is based on a HMAC (Hash-based Message Authentication Code) scheme to encrypt data two times: one time when the file is uploaded and the other when the file is distributed. Using two times encryption, the computation complexity time of the encryption method is increased. In [7], an ASIF algorithm is proposed, which is a hybrid encryption model in which data are compressed and then encrypted when the user wants to upload his/her data to the server; in this manner, the model will save space on the server and reduce link and channel congestion. In this algorithm two keys (k1, reverse key) are generated; then a random number is calculated that represents the position of character in k1; this random number is added to k1 and then XOR is applied on it; then each resultant value is converted to its correspondent character in ASCII code that will generate k2, then the plain text is XORed with k1 or k2. The length of k1 is equal to 72 bits. The decryption algorithm performs the reverse order of operation of the encryption algorithm. The results presented in the paper conclusion indicate that the algorithm provides better security. In [8], a multilevel encryption algorithm is proposed, which uses in the first level the AES algorithm and in the second level the rounded shifted algorithm that is modified Caesar algorithm. When the user uploads the file, it is first encrypted using AES and then the resultant cipher text will be encrypted using Caesar algorithm and stored encrypted on the cloud database. For retrieving the data, they are decrypted on the server in the reverse order of encryption and the original data are sent to the server. In [9], a multilevel encryption technique is proposed, which uses the AES algorithm for the text encryption and elliptic curve cryptography to encrypt the key during transmission over the network. In [10], a multilevel encryption/decryption algorithm is proposed, in which the DES algorithm is used as the first-level encryption when the user uploads the file to the server and then the RSA algorithm is used as a second encryption algorithm, and then data are stored on the database encrypted. For data retrieval, data are decrypted in the reverse manner as encryption on the server and sent in plain text to the user. In this scenario, data are travelled in plain text send-receive in the network.

In [11], an algorithm is proposed, in which a transposition cipher is used first on the plain text and then Caesar cipher is applied to obtain the final encrypted text. The transposition will move down the letters in the odd location and then add them at the end of the text.

Example: Summer will be divided as follows:
Higher level: u m r
Low level: S m e
Final text to be encrypted: umrSme

Then the final text will be encrypted using Caesar algorithm. In [12], the implementation of three-level encryption algorithm that uses RSA, random number generator, and 3DES to ensure data security is proposed. The algorithm is implemented in C# using SQL Azure service database in which the user can choose the order of the algorithm they want. In [13], the authors implemented the work presented in [5] to be applied on DBMS; first, data were encrypted using DES and then encrypted using the RSA algorithm. The results show that if an intruder accesses the data and intercepts the data, and decrypts them, accessing the original data will be difficult. In [14], an algorithm for maintaining the integrity of data on the cloud server is proposed, in a manner that data are encrypted on the server side and decrypted on the client side. The main basic idea is to use for any character a key different to encrypt and decrypt. The proposed algorithm divides the text file into blocks, the binary equivalent of blocks is stored in an array, then these data are circularly rotated, and encryption in any rotation is divided the data by 2 to ensure the privacy of data. That means if the file is composed of n blocks with each block having n characters, the number of operations performed is $n*n$ and the complexity of the algorithm is O $(n*n)$. The performance analysis presented in the paper indicated that the proposed encryption/decryption algorithm maintained the security of data without increasing storage space and overhead computation. In [15], to maintain the data security in cloud computing, the homomorphic encryption system that makes an operation on data without decryption is proposed. The idea is to store data encrypted and to retrieve them encrypted; in this manner, the holder of the private key is only the sender and only he can read or work on plain text. The cloud server make operations on and store only the encrypted data.

8.3 ALGORITHM DESCRIPTION

Our algorithm works in phases. In the first and the second phases, the substitution method to encrypt the data with two different keys and two different methods is used, and in the third phase, the resultant text will be converted to binary and XORed with a new key and then stored in a binary or hexadecimal number.

8.3.1 Encryption algorithm

Consider M represents the plain text, E represents the encrypted text, and K represents the key.

First phase

- Insert the plain text M.
- Generate a random key $k1$ with a length of 10–30 characters; the key can be composed of small and capital letters and the numbers from 0 to 9.
- Prepare an 8*8 matrix, in which in the first location (1*1) @ character and in the last location (8*8) the # character are inserted. Then, the key characters one by one by without repetition is inserted, where in each cell only one character is inserted. Then complete the matrix with the remaining characters and numbers that are not presented in the key. To generate the first encryption text $E1$, follow the following rules.

8.3.1.1 Encryption rules

Divide the plain text M into different parts, each one composed of only two characters. Then, each two characters are encrypted together according to the following rules:

1. If the two characters are located on the same row, the encryption of each character is its consecutive one from the left to the right.
2. If the two characters are on the same column, the encryption of each character is the consecutive one from top to bottom.
3. If the two characters are located in two different columns or two different rows, the encryption is the cross.
4. To encrypt the last character in the row, move in counterclockwise.

5. To encrypt the last character in the column, move in a circular manner from the bottom to top.
6. The consecutive repeated characters are separated by the @ character and encrypted following the rules from 1 to 5.
7. If the size of the text is odd, add at the end the @ character and encrypt using the rules from 1 to 5.

8.3.1.2 Second phase

Generate $k2$ that is a random number between 2 and 63, and calculate the encryption text using the following equation:

$$E2 = (E1 \oplus k2)\bmod 63 \tag{8.1}$$

8.3.1.3 Third phase

Convert $E2$ to hexadecimal, then binary, and store it into $E3$. Generate a new hexadecimal key $k3$, convert to binary, and then apply the following equation for encryption:

$$E = E3 \oplus k3 \tag{8.2}$$

8.3.1.4 Encryption example

In this example, we will describe in simple mode how the algorithm will work taking a fixed key value, but in the implemented system, the three keys are selected randomly. That means encrypting the file different times will be encrypted by different keys, but each time, the key is stored to be used in the decryption phase:

M = extend, k = Sandra09
Division phase: ex te nd
Encryption matrix:

@¤	S¤	a¤	n¤	d¤	r¤	0¤	9¤	¤
A¤	B¤	C¤	D¤	E¤	F¤	G¤	H¤	¤
I¤	J¤	K¤	L¤	M¤	N¤	O¤	P¤	¤
Q¤	R¤	T¤	U¤	V¤	W¤	X¤	Y¤	¤
Z¤	b¤	c¤	e¤	f¤	g¤	h¤	i¤	¤
j¤	k¤	l¤	M¤	o¤	p¤	q¤	s¤	¤
t¤	u¤	v¤	W¤	x¤	y¤	z¤	1¤	¤
2¤	3¤	4¤	5¤	6¤	7¤	8¤	#¤	¤

Encryption:

ex	Te	Nd
wf	Zw	Dr

So, the encrypted text in the first phase $E1$ = "wf Zwdr".

E1	w	F	Z	w	D	r
E2=E1+26	U	E	0	U	C	Q

In the second encryption phase, consider the $K2=26$.
So, $E2$=UE0UCQ.
In the third encryption phase, $E2$ is converted to hexadecimal.

E2	U	E	0	U	C	Q
Hexa	55	45	30	55	43	51
Binary	01010101	01000101	00110000	01010101	01000011	01010001
⊕						
K3	123BF9CAB75=00000001001000111011111111001110010101011011 0101					
E	01010100	01100110	10001111	11001001	11101000	00100100

A complete scenario of encryption and decryption processes is presented in simulation results Section 8.4.1.

8.3.2 Decryption Algorithm

As presented in the encryption algorithm, the final encrypted text is in binary or in hexadecimal that is considered the input to the decryption algorithm. Considering that we use symmetric encryption, one key is used for encryption and decryption in each phase. The final encrypted text is decrypted three times in reverse manner to arrive at the plain text.

The first decryption phase consists in using exclusive-or between the key and the final encrypted text E according to the following rules: from equation (8.2) $E=E3 \oplus k3$ making exclusive-or for the key $k3$ to the two parts of equation (8.2), will produce the following: $k3 \oplus E=E2 \oplus k3 \oplus k3=E2$ that means knowing E and $k3$ will turn to $E2$.

Second decryption phase: $E1 = (E2 - k2)$ mod 63.

Third decryption phase: To obtain the plain text, the decryption is the reverse process that follows the same rules applied on the same matrix used in the encryption process, but in the reverse order.

E	01010100	01100110	10001111	11001001	11101000	00100100
⊕						
K3	00000001	00100011	10111111	10011100	10101011	01110101
E2	01010101	01000101	00110000	01010101	01000011	01010001
Hexa	55	45	30	55	43	51
Char (E2)	U	E	0	U	C	Q
E1 = E2–26	w	F	Z	W	d	r
Go to matrix						
M	e	X	t	E	n	d

Decryption example:

Decryption matrix

Decryption matrix								
@□	S□	a□	n□	d□	r□	0□	9□	¤
A□	B□	C□	D□	E□	F□	G□	H□	¤
I□	J□	K□	L□	M□	N□	O□	P□	¤
Q□	R□	T□	U□	V□	W□	X□	Y□	¤
Z□	b□	c□	e□	f□	g□	h□	I□	¤
J□	k□	l□	m□	O□	p□	q□	S□	¤
T□	u□	v□	w□	X□	y□	z□	1□	¤
2□	3□	4□	5□	6□	7□	8□	#□	¤

8.3.2.1 Algorithm analysis

The main idea behind our work was to maintain the data confidentiality and integrity, which means data can be accessed or changed only by the authorized persons; for that, we apply the encryption in levels to ensure that no one can read the plain text without having many information that cannot be obtained in an easy manner. In the first level of encryption, data are encrypted and sent to the cloud server. That means data travel in network encrypted; if an intruder tries to access the data, he/she must know the key and the encryption method; our key is of length 10–30 characters; for that, the number of attempts to crack the key must be in the degree of 1,064–3,064 considering capital and small letters and the numbers from 0 to 9. In addition, our matrix is extended to 64 cells. For that, accessing the data cannot be easy for any one. The second-level encryption is performed on the server, the number of attempts to crack the second key is on the range of 63, but still the original data are hidden. The last encryption step is performed before storing the data in the cloud database and the number of attempts to crack the key is 1,616, where the key is of length 16 characters in the hexadecimal system. While we speak about cloud, we cannot speak about absolute security; the possibility of attacks exists all the time. In our work, we have made the possibility to read the original text more complex. Our algorithm is easy to implement and does not require complex computation, it is easy to encrypt any text, but it is not easy to decrypt it without knowing the needed information.

8.4 SIMULATION AND PERFORMANCE ANALYSIS

Our system is implemented using visual C#, in which our algorithm is implemented in its three phases. In the first phase, the function that represents the encryption rules on the matrix of size 8*8 is implemented. In the second phase, the substitution method that uses a key from 2 to 63 is implemented and a random function (next in class random in.net) that generates the key randomly is used. In the third phase, is used the function "texttobinary" to convert the resultant encrypted text to binary and then apply to it the exclusive-or operator with the key that is generated randomly using the function "atmkeygenerator". Figures 8.2–8.4 show an example on the algorithm implementation (Figure 8.5).

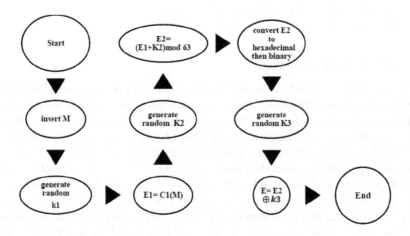

Figure 8.2 Encryption algorithm description.

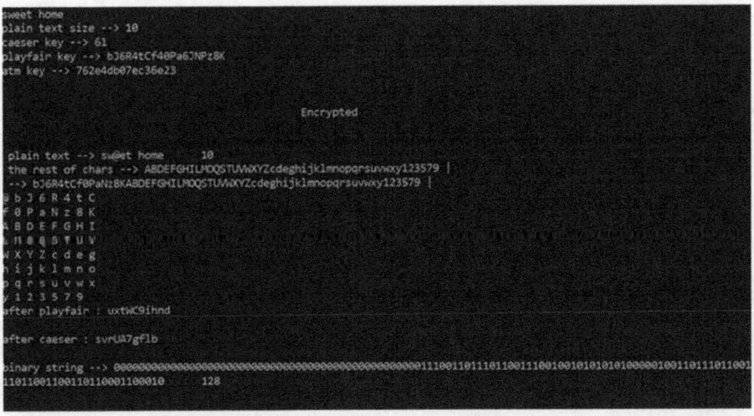

Figure 8.3 The encryption process of the text sweet home.

Figure 8.4 The decryption process.

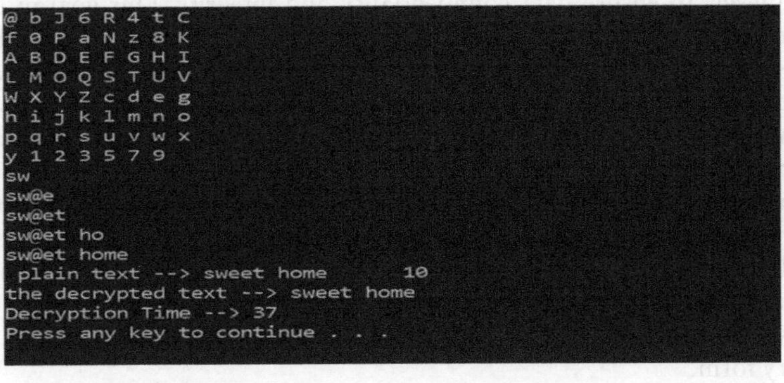

Figure 8.5 Plain text recovery.

Table 8.1 Performance analysis

Text length (character)	k1 length (character)	k2 length (int)	k3 (bit)	Encryption time (ms)	Decryption time (ms)
50	30	10	64	13	22
100	27	31	64	26	39
150	23	50	64	38	53
200	14	40	64	47	72
250	28	13	64	60	92

8.4.1 Performance analysis

Cloud security is an important challenge that poses many problems on the efficiency of cloud services; for that, researchers pay attention on working in this field. Our paper is the first one that implements and concatenates three symmetric methods to obtain best results, most of the works are not implemented, but only proposed, and some other works consist in simple comparison between algorithms such as DES, AES, and Blowfish. To measure the performance of our work, we use the meantime parameter. Meantime means the difference between the starting and ending times of encryption taken for a particular algorithm. Our results show that if the text size is increased, the encryption time will increase and the decryption time is increased, which is compatible with the results presented in [16] as mentioned in Table 8.1. The best encryption algorithm is easy to encrypt, but difficult to decrypt.

8.5 CONCLUSIONS AND FUTURE WORK

We have proposed a new encryption method that applies encryption in three levels to ensure data confidentiality and integrity. Our goal in this work was to distract the attention of the attackers from our data while they are in transmission over the Internet or on the server. We have made the comprehension of the data difficult even if someone has intercepted them successfully, because knowing the explicit text needs to know many information such as three keys generated randomly and three encryption methods. Our algorithm is easy to implement, it does not need complex computation for encryption or decryption, and the time needed to encrypt the data is small, while decrypting the data requires more time as presented in our results. Our future work will be working on binary file to make the decryption more difficult to return from the encrypted text to its original binary form.

Compliance with Ethical Standards:
This research was funded by Jadara University, Jordan, as a part of Scientific Research Fund in the University. This research is ethically approved, and it does not contain any studies with human participants or animals performed by any of the authors.

REFERENCES

1. Wahid, M.N.A., Ali, A., Esparham, B. and Marwan, M. "A comparison of cryptographic algorithms: DES, 3DES, AES, RSA, and blowfish for guessing attacks penetration", (2018, August). *Journal of Computer Science applications and Information Technology* Vol. 3, No. 2, 1–7. Doi: 10-15226/2474-9257/3/2.
2. Ahmet, E.F.E. and ISSI, H.N. "Cryptography challenges of cloud computing for E- government services", (2018). *International Journal of Innovation Engineering Applications* Vol. 2, No. 2, 4–14.
3. Yerlikaya, T., Bulus, E. and Arda, D. "Eliptik Egui Sifreleme Agorithmasi Kullanan Dijital Imza Uygulamasi", (2018). Researchgate 2018. http://g00.gt/mheegl.
4. Kaur, M. and Singh, R. "Implementaion encryption algorithms to enhance data security of cloud computing", (2013, May). *International Journal of Computer Applications* Vol. 70, No. 18, 16–21.
5. Morris, L. "Analysis of partially and fully homomorphic encryption", (2013). Available at the following link: www.semanticscholar.org/paper/analysis.
6. Rahul, G. and Subashini, N.J. "Secure storage services and erasure code implementation in cloud servers", (2014, January). *International Journal of Engineering Research of Technology (IJERT)* Vol. 3, No. 1, 1810–1814.
7. Mushtaque, M.A., Dhiman, H. and Hussain, S. "A hybrid approach and implementation of a new encryption algorithm for data security in cloud computing", (2014). *International Journal of Electronic and Electrical Engineering* Vol. 7, No. 7, 669–675.
8. Rai, D., Desai, R., Tripti, P.S. and Vinutha, B. "Multilevel encryption for cloud storage", (2018). *SAHYADRI International Journal of Research* Vol. 4, No. 1, 2018. http://www.sijir.in/.Journal.
9. Jana, B., Poray, J., Mandal, T. and Kule, M. "A multilevel encryption technique in cloud security", (2017). *The 7th International Conference on Communication Systems and Network Technologies (CSNT).* Doi: 10.1109CSNT.2017.8718541, Naqpur, India.
10. Satyanarayana, K. "Multilevel security for cloud computing using cryptography", (2016, February). *International Journal of Advanced Research in Computer Engineering & Technology (IJARCET)* Vol. 5, No. 2, 17338–17346.
11. Subharsi, P., and Pandmapriya, A. "Multilevel encryption for ensuring public cloud", (2013, July). *International Journal of Advanced Research in Computer Science and Engineering* Vol. 3, No. 7, www.ijarcsse.com.

12. Kaur, A. and Bhardwaj, M. "Hybrid encryption for cloud database security", (2012). *International Journal of Engineering Science & Advanced Technology* Vol. 2, No. 3, 737–741.

13. Khan, S.S. and Tuteja, R.R. "Cloud security using multilevel encryption algorithms", (2016, January). *International Journal of Advanced Research in Computer and Communication Engineering* Vol. 5, No. 1, Doi: 10.17148IJRCCE.2016.5116.

14. Prakash, G.L., Prateek, M. and Singh, I. "Data encdyption and decryption algorithm using key rotation for data security in cloud system", (2014, April). *International Journal of Engineering and Computer Science* Vol. 3, No. 4, 5215–5223.

15. Tebaa, M. and Hajji, S.E. "Secure cloud computing through homomorphic encryption", (2013, December). *International Journal of Advancement in Computing Technology (IJACT)* Vol. 5, No. 16, 29–38.

16. Mewada, S., Shrivastava, A., Sharma, P., Purohit, N. and Gautam, S.S. "Performance analysis of encryption algorithms in cloud computing", (2016). *International Journal of Computer Science and Engineering* Vol. 3, No. 2, 83–89.

Chapter 9

Anomaly detection using a class of supervised and unsupervised learning algorithms

Bhaumik Joshi, Devang Goswami,
and Harsh S. Dhiman
Adani Institute of Infrastructure Engineering

CONTENTS

9.1 INTRODUCTION

With the rise of the digital age, almost everything is monitored via the Internet, for example the number of taxis booked in a day, the output of any instrument, and the traffic of a network. With this, there has been an exponential surge in the time series dataset. However, it is very important to detect the anomaly and check if the system is healthy and performing inbound limits. Whenever an unseen abnormality or behaviour change occurs at the output of the system, we consider it to be an anomaly. Anomaly detection means detecting unseen abnormalities in the time step data (Agrawal and Agrawal 2015; Ramchandran and Sangaiah 2018).

Irregularity identification is an integral issue that has been studied for many decades, and from that, many kinds of literature, algorithms, and thesis were proposed (Shipmon et al. 2017). Many outlier identification methods have explicitly been produced for certain application areas, while others are more general. The necessity of accurate intrusion identification is because failing to detect one can result in a severe loss and sometimes cause severe accidents in the system.

Distinguishing abnormal activities in the information is being studied as early as the 19th century. One cannot justify the noise in the system as an anomaly as there is no parity between the noise and an anomaly. Anomaly and noise are a completely different phenomenon. Anomaly identification systems are different from noise removal or noise accommodation, as these deal with the noise in our dataset. Noise is an unwanted abnormality occurring frequently, which is an obstruction for the data analytics for the analysis of the dataset. So, it is necessary to remove noise before we begin our data analysis.

Irregularity identification was suggested for intrusion detection systems (IDS) by Dorothy Denning in 1986. Intrusion identification for IDS is generally identified with benchmarks and measurements. However, intrusion detection can also be performed with inductive learning and soft computing. Different types of methods introduced in 1999 included remote hosts, profiles of users, networks, groups of users, workstations, and programs based on means, frequencies, variances, standard deviations, and co-variances. The partner of irregularity identification in interruption recognition is abuse location.

Since the identification of these anomalies is crucial, there had been a significant development in locating these outliers and many algorithms were proposed to identify these anomalies during the past decade. Machine learning is used to automatize anomaly detection. Anomaly detection methods are classified into two primary methods: supervised methods (Dhiman, Anand, and Deb 2018; Dhiman et al. 2020; Dhiman, Deb, and Guerrero 2019) and unsupervised methods (Caruso and Malerba 2007).

1. Supervised methods: They require a labelled dataset as anomalous and normal. They require training data for anomaly detection. Bayesian network, k-NN (k-nearest neighbours) method, and neural networks are examples of supervised methods of anomaly detection.
2. Unsupervised methods: They assume statistically from most of the datasets the data are normal and less amount of data is anomalous. They do not require any pre-labelled dataset and training data. Examples of unsupervised methods for anomaly detection are hypothesis-based analysis, auto-encoders, and the k-means method. The scope of anomaly detection is vast in deep learning, machine learning, and artificial intelligence with increasing data.

This chapter compares five algorithms: support vector machine, artificial neural network, long short-term memory network, isolation forest, and K-means on four time series datasets for the identification of anomalous data points. Later, they find the absolute error between the predictions made by SVM, ANN, and LSTM with the anomalous data points in the dataset. We used the adaptive threshold for detecting the anomaly in the actual dataset and later comparing anomalous data points with K-means clustering and isolation forest algorithms.

The rest of the chapter is divided into six sections: In Section 9.2, we discuss the adaptive threshold and various regression techniques such as SVM, ANN, and LSTM. In Section 9.3, we discuss the unsupervised learning algorithms, for instance isolation forest and K-means clustering. In Section 9.4, an overview of the open-source datasets used for this research is elaborated. In Section 9.5, we discuss the empirical results obtained through adaptive threshold, isolation forest, and K-means clustering. Lastly, in Section 9.6, we sum up the outcomes and conclude the best method possible for anomaly detection.

9.2 ADAPTIVE THRESHOLD AND REGRESSION TECHNIQUES FOR ANOMALY DETECTION

Anomaly detection by shifting through different feature subsets has high efficiency and is effective in many cases. However, the identification of relevant feature anomalies is non-trivial and difficult (Chandola, Banerjee and Kumar 2009). It comprises of recognizing irregular data points by distinguishing outliers from the normal datapoints in the dataset, which is normal and not an anomaly. An effective intrusion identification method has to adapt the identification process for each system condition and each time series behaviour (Dani et al. 2015). In this method, we check if the value lies between the summation of mean and two times the standard deviation, and the subtraction of mean and two times the standard deviation of the dataset column, as shown below. If the value lies outside the boundaries, we will mark it as an anomaly or anomaly behaviour of the system. We found the absolute error in the points that were detected as an anomaly using this method and the predictions made at the same time. Given a univariate time series X_i, $i = 1, 2, ..., N$, the adaptive threshold limits can be expressed as

$$\mu - 2\sigma \leq X_i \leq \mu + 2\sigma \tag{9.1}$$

where μ and σ are the mean and standard deviation for the time series, respectively.

9.2.1 Artificial neural networks

Artificial neural network, the phenomenal idea was derived from one of the complex things in this universe, the brain. Artificial neural networks have three layers, mainly recognized as the input, hidden, and output layers. Each layer has a specific number of neurons interconnected with the previous and the next layers, respectively. Each neuron consists of weights and bias. For high-accuracy predictions, the network should be having optimal weights. The input layer is in charge of feeding the data to the system; it is the first layer in the ANN. The layers in the middle of the input layer and the output layer are known as the hidden layers. The number of hidden layers is decided based on the complexity of the dataset. The output layer predicts the output, as our problem is based on the continuous regression-based predictions; the network has only one neuron in its output layer. The ANN uses a loss function called mean squared error for calculating the cost of the neural network, for continuous regression.

9.2.2 Long short-term memory recurrent neural network

Long short-term memory recurrent neural networks contain cyclic connections that make them more proficient than an ordinary feed-forward neural network (Malhotra et al. 2016). The LSTM contains a special block known as the memory blocking the recurrent hidden layer. These blocks contain memory cells capable of storing the time-related condition of the system in addition to a special unit called gates to control the progression of data. Each memory block has an input and an output gate. The input gate runs the information stream enactment into the memory cell. The output gate runs the output information stream enactment towards the remainder of the LSTM. The forget gate entryway scales the inside condition of the cell before contributing it as a contribution to the cell through the self-intermittent association of the cell, thereby adaptively overlooking or resetting the cells' memory. LSTM networks are appropriate for classification, processing, and generating empirical results based on time series data since there can be slacks of the obscure term between significant occasions in time arrangement information (Sakurada and Yairi 2014).

9.3 UNSUPERVISED LEARNING TECHNIQUES FOR ANOMALY DETECTION

9.3.1 Isolation forest

The word isolation represents separating one or more data points from the entire dataset. This is a recursive algorithm and is called on the dataset; the partitions are created in the dataset until all the data points are isolated. This partition of the dataset is done randomly, this partition implies a shorter path in the tree for anomalies, and with fewer data points, partitions would be fewer. Also, the data points that are distinguishable are partitioned earlier. Hence, when the tree generates shorter path lengths for a data point, then the chance of that point being an anomaly increases, compared to other data points in the tree. The isolation tree can be represented as, assuming T being node of the isolation tree. T node can be either a leaf node with no children, or an internal node with one test and precisely two daughter nodes represented as (Tl, Tr). A test contains an attribute q and a split value p such that $q < p$, which divides the instance into Tl and Tr. This is also a promising algorithm when it comes to the identification of intrusions (Liu, Ting, and Zhou 2008).

9.3.2 K-means clustering

This unsupervised algorithm divides the dataset into k subgroups called clusters represented as $C1, C2,..., Ck$. Here, k is a predefined parameter. The algorithm is used in practice for pattern recognition, clustering analysis, and anomaly detection (Karami and Guerrero-Zapata 2015). The clusters are created based on the minimization of the sum of squares of distances between data and the corresponding subgroup. Finding the perfect match of k for a given problem is quite difficult; for a dataset with n number of data points and d dimensions, k could be discovered using the trial-and-error method. The number of iterations for this trial-and-error method can vary in a wide range from some to many thousands, and the number of iterations is directly proportional to the number of groups, input data distribution, and a number of patterns. Therefore, the direct implementation of this algorithm can be very costly.

We have used a wide range of k for the dataset; later, we have used the elbow method to find the optimal value of k for the dataset. The selection of optimal k using the elbow method is by running the algorithm on different values of k, later plotting the graph of cluster score. The score is generally some form of intra-cluster distance relative to inter-cluster distance. The number of clusters was chosen where the difference between scores is smaller than the 90 percentile if the elbow curve is hard to define. The scikit-learn also provides an algorithm called silhouette score for determining the optimal k; however, the computational power increases with the use of this algorithm, and no significant difference is found while using elbow over the silhouette score.

9.4 DESCRIPTION OF THE DATASET

We have worked with four different time series datasets. The first dataset is an open-source dataset available on Kaggle, which is named "HOUSEHOLD ELECTRIC POWER CONSUMPTION". The dataset is a multivariate time series dataset from Kaggle (n.d.). The dataset contains 2,075,259 data points gathered from December 2006 to November 2010. The dataset carries the date, household global minute-averaged active power (in kW), time, voltage (in volt), reactive power (in kW), and current intensity (in ampere). There are three sub-metering, and all are in watt-hour of active energy. Energy sub-metering No. 1: It relates to containing mainly a dishwasher, kitchen, a microwave, and an oven. Energy sub-metering No. 2: It relates to containing a laundry room, a tumble drier, a light, and a refrigerator. Energy sub-metering No. 3: It corresponds to an air conditioner and an electric water heater.

The second and third datasets were open-source on GitHub. The datasets are the collections of Twitter mentions of huge traded stocks on an open market organization, for example Apple and Google (Lavin n.d., 2015). The metric value represents the number of indications for a given ticker symbol every 5 minutes. The datasets D2 and D3 contains 15,903, 15,843 datapoints respectively. The first column consists of the timestamps, and the second column consists of the values recorded. The fourth dataset was open source and also available on GitHub. The number of New York City taxi passengers dataset, where five outliers arise during the event of Thanksgiving, the New York City Marathon, Christmas, a snowstorm, and New Year's Day. The data file included here consists of aggregating the total number of taxi passengers into 30-minute buckets. The dataset contains 10,321 rows. The first column is of timestamps, and the second column contains the values.

We have processed the dataset and replaced the missing entries with the mean of the respective column, later standardizing the features by subtracting the mean and then scaling to the unit variance. The dataset has been divided into three categories: 60% of the dataset is for training the models, 20% of the dataset is for validation, and the rest 20% is for testing the model. Figure 9.1 represents the flow for anomaly detection (Table 9.1).

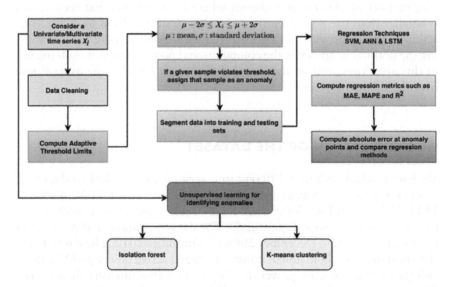

Figure 9.1 Flowchart for anomaly detection using adaptive threshold and unsupervised learning.

Table 9.1 Different datasets for anomaly detection

Dataset	Description	Total size	Training/validation/ testing
D1	2006-12-16 17:24:00 to 2010-11-26 21:02:00	2,075,259	1245,154/415,050/ 415,055
D2	2015-02-26 21:42:53 to 2015-04-23 02:47:53	15,903	9,540/3,180/3,183
D3	2015-02-26 21:42:53 to 2015-04-22 21:47:53	15,843	9,505/3,169/3,169
D4	2014-07-01 00:00:00 to 2015-01-31 23:30:00	10,321	6,192/2,064/2,065

9.5 RESULTS AND DISCUSSION

The present study on anomaly detection was performed on Google Colaboratory on four different open-source datasets. For SVM, isolation forest, and K-means, we used the prebuild algorithm proposed by the scikit-learn library. For SVM, the radial basis function (RBF) kernel was used, with regularization parameter as 1 and epsilon as 0.2. As the contamination rate was not known for all the datasets, it was kept to auto and n estimators were kept in 1,000. We used the Keras library for this project. A high-level library built on top of TensorFlow, Keras provides some finest application programming interfaces (APIs), which ease the development process of creating a deep learning network. We also used Pandas, NumPy, and Matplotlib for data manipulation and plotting the graphs. For each dataset, different ANN and LSTM networks were created, and with the hyper-parameter optimization step, we found the optimum parameters for the respective datasets. The results were predicted by different networks. Figure 9.2 illustrates the estimations by tested models for dataset D1.

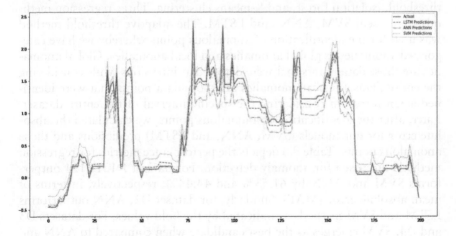

Figure 9.2 Prediction using SVM, ANN, and LSTM for dataset D1.

Table 9.2 Number of anomalies identified by different methods

Datasets	Adaptive threshold		Isolation forest	K-Means clustering
	Global	Local		
D1	18,263	27,618	206,787	65,518
D2	28	151	1,813	742
D3	150	148	3,080	327
D4	6	37	5,109	1,374

Table 9.3 Performance metrics of SVM, ANN, and LSTM on different datasets

Dataset	Model	MAE	MSE	R2 score	Theil's U1	Theil's U2
D1	SVM	0.13	0.01	0.97	0.16	0.08
	ANN	0.09	0.02	0.96	0.18	0.1
	LSTM	0.05	0.005	0.99	0.09	0.04
D2	SVM	0.15	1.17	0.47	0.72	0.50
	ANN	0.13	0.47	0.78	0.46	0.25
	LSTM	0.14	1.13	0.48	0.71	0.51
D3	SVM	0.15	0.02	0.96	0.18	0.08
	ANN	0.37	0.37	0.51	0.69	0.43
	LSTM	0.38	0.38	0.49	0.70	0.45
D4	SVM	0.10	0.01	0.98	0.12	0.06
	ANN	0.17	0.05	0.95	0.21	0.11
	LSTM	0.17	0.05	0.94	0.22	0.11

Table 9.2 shows the number of anomalous points identified by adaptive threshold, isolation forest, and K-means clustering. Three regression methods were used: SVM, ANN, and LSTM. The adaptive threshold method was used for the identification of anomalous points whereby we have categorized anomalies as global anomalies and local anomalies. Global anomalies are those data points that were detected as intrusions while considering the entire dataset. Local anomalies are those data points that were identified as an intrusion considering the specific interval of the entire dataset. Later, after the identification of anomalous points, we calculated the absolute error for our models (SVM, ANN, and LSTM) predictions and those anomalous points. Table 9.3 depicts the performance metrics for regression methods discussed for anomaly detection. For dataset D1, LSTM outperforms SVM and ANN by 61.53% and 44.44%, respectively, in terms of mean absolute error (MAE). Similarly, for dataset D2, ANN outperforms LSTM and SVM methods as indicated by the bold values. For datasets D3 and D4, SVM emerges as the best candidate when compared to ANN and LSTM.

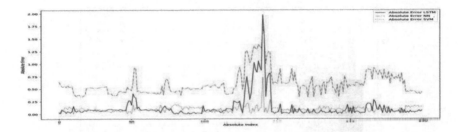

Figure 9.3 Absolute error at various anomaly indices using LSTM, ANN, and SVM for household electric power consumption.

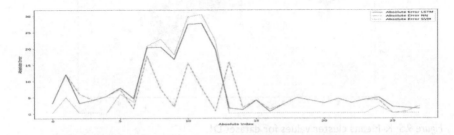

Figure 9.4 Absolute error at various anomaly indices using LSTM, ANN, and SVM for Twitter mentions of Apple.

Figures 9.3 and 9.4 represent the absolute difference between the estimations by various regression techniques and the anomalous data points. The more the absolute error, the higher the chances are for a point to be an anomaly. The spikes in the figure explicitly represent that there is a notable deviation between the model's estimation and the true value at the given data point. According to the dataset, a particular absolute error threshold could be calculated, and when the absolute error of the actual value and our model's prediction crosses the absolute error threshold, an alarm should raise calling that data point an anomaly. For the adaptive threshold, the anomalies were recorded on the test dataset only. For unsupervised learning, two algorithms were used in isolation forest and K-means clustering. The anomalies were recorded on the entire dataset. The isolation forest algorithm was trained on the dataset, and when tested on the dataset, it categorizes the predictions into two parts: The algorithm outputs either 1 or −1, where 1 represents the data point to be normal and −1 represents the data point to be an anomaly. Table 9.2 represents the number of anomalies detected by the isolation forest algorithm on the respective datasets.

Figure 9.5 shows the four clusters out of 17 clusters for the D1 dataset. For every cluster, the radius P is calculated; if any data point falls out of the

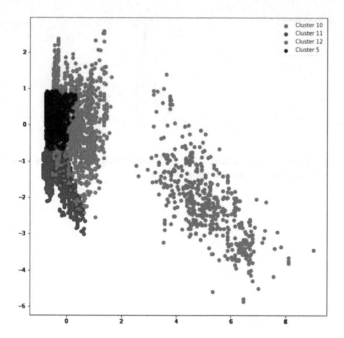

Figure 9.5 K-Means cluster values for dataset D1.

radius *P*, that point would be labelled as an anomaly. However, the above figure explicitly implies that many data points in red had formed a cluster, which is far away from the normal data points. Hence, one could justify that the data points in this cluster are the anomalous data points without any *P* radius.

9.6 CONCLUSIONS

In this work, we discuss five algorithms: SVM, ANN, LSTM, isolation forest, and K-means clustering on four open-source datasets. SVM, ANN, and LSTM were used for the continuous regression estimation. With the use of the adaptive threshold method, anomalous points were recorded and the absolute error was calculated among the estimations and the true anomalous points. Based upon the dataset, an absolute error threshold can be decided for anomaly identification. In Kaggle (2017), it has been stated that the dataset contains five anomalies in Table 9.2. We can see that the adaptive threshold method has identified six global anomalies. Also, the method has a setback; if the dataset shows a surge or a fall at every particular time interval, that rise or dip cannot be considered an anomaly; in this case, the adaptive threshold will give a false positive. The isolation forest, a promising algorithm for anomaly detection, only applies if the data are volatile;

for the dataset D4, the isolation forest has recorded more than 5,000 anomalous points. *K*-Means clustering is a technique that one can use for the detection of an anomaly. *K*-Means does not hold back for the above two cases; however, the impeccable calculation of radius *P* is required for a cluster; if any point falls out of the radius *P* of the particular cluster, then it can be labelled as an anomalous data point. It is always better to have false positives and rectify them than to have a false negative.

REFERENCES

Agrawal, S., and J. Agrawal. 2015. "Survey on anomaly detection using data mining techniques." *Procedia Computer Science* 60: 708–713. Doi: 10.1016/j.procs.2015.08.220.

Caruso, C., and D. Malerba. 2007. "A data mining methodology for anomaly detection in network data." In *Lecture Notes in Computer Science*, 109–116. Springer, Berlin Heidelberg. Doi: 10.1007/978-3-540-74827-4_14.

Chandola, V., A. Banerjee, and V. Kumar. 2009. "Anomaly detection." *ACM Computing Surveys* 41 (3): 1–58. Doi: 10.1145/1541880.1541882.

Dani, M.-C., F.-X. Jollois, M. Nadif, and C. Freixo. 2015. "Adaptive threshold for anomaly detection using time series segmentation." In *International Conference on Neural Information Processing*, 82–89. Springer, Istanbul.

Dhiman, H.S., P. Anand, and D. Deb. 2018. "Wavelet transform and variants of SVR with application in wind forecasting." In *Advances in Intelligent Systems and Computing*, 501–511. Springer, Singapore. Doi: 10.1007/978-981-13-1966-2_45.

Dhiman, H.S., D. Deb, J. Carroll, V. Muresan, and M.-L. Unguresan. 2020. "Wind turbine gearbox condition monitoring based on class of support vector regression models and residual analysis." *Sensors* 20 (23): 6742. Doi: 10. 3390/s20236742.

Dhiman, H.S., D. Deb, and J.M. Guerrero. 2019. "Hybrid machine intelligent SVR variants for wind forecasting and ramp events." *Renewable and Sustainable Energy Reviews* 108: 369–379.

Kaggle n.d. "Household electric power consumption — Kaggle." https://www.kaggle.com/uciml/electric-power-consumption-data-set. (accessed: 15 October 2020).

Kaggle. 2017. "New York city taxi trip duration — Kaggle." https://www.kaggle.com/c/nyc-taxi-trip-duration, September. (accessed: 15 October 2020).

Karami, A., and M. Guerrero-Zapata. 2015. "A fuzzy anomaly detection system based on hybrid PSO-Kmeans algorithm in content-centric networks." *Neurocomputing* 149: 1253–1269. Doi: 10.1016/j.neucom.2014.08.070.

Lavin, A. 2015. "NAB/Twitter volume AAPL.csv at master numenta/NAB GitHub." https://github.com/numenta/NAB/blob/master/data/realTweets/Twitter_volume_AAPL.csv, Feb. (accessed: 17 October 2020).

Lavin, A. n.d. "NAB/Twitter volume GOOG.csv at master numenta/NAB GitHub." https://github.com/numenta/NAB/blob/master/data/realTweets/Twitter_volume_GOOG.csv. (accessed: 17 October 2020).

Liu, F.T., Ting, K.M., and Zhou, Z.-H. 2008. "Isolation forest." In *2008 Eighth IEEE International Conference on Data Mining*, 413–422. IEEE, The University of Cadiz, Spain.

Malhotra, P., A. Ramakrishnan, G. Anand, L. Vig, P. Agarwal, and G.M. Shroff. 2016. "LSTM-based encoder-decoder for multi-sensor anomaly detection." CoRR abs/1607.00148. http://arxiv.org/abs/1607.00148.

Ramchandran, A., and A.K. Sangaiah. 2018. "Unsupervised anomaly detection for high dimensional data—An exploratory analysis." In *Computational Intelligence for Multimedia Big Data on the Cloud with Engineering Applications*, 233–251. Elsevier. Doi: 10.1016/b978-0-12-813314-9.00011-6.

Sakurada, M., and T. Yairi. 2014. "Anomaly detection using autoencoders with nonlinear dimensionality reduction." In *Proceedings of the MLSDA 2014 2nd Workshop on Machine Learning for Sensory Data Analysis - MLSDA'14*, ACM Press. Doi: 10.1145/2689746.2689747.

Shipmon, D., J. Gurevitch, P.M Piselli, and S. Edwards. 2017. *Time Series Anomaly Detection: Detection of Anomalous Drops with Limited Features and Sparse Examples in Noisy Periodic Data*. Technical Report. Google Inc. https://arxiv.org/abs/ 1708.03665.

Chapter 10

Improving support vector machine accuracy with Shogun's multiple kernel learning

Debani Prasad Mishra and Pallavi Dash
International Institute of Information Technology Bhubaneswar

CONTENTS

10.1 INTRODUCTION

The support vector machine (SVM) can be classically viewed as perceptron augmentation. The minimization of misclassification errors is done by using the perceptron algorithm. In SVMs, however, our aim for optimization is to maximize the margin. Although SVM enjoys high popularity in classifying linear separable data, it is commonly used for classifying data that are non-linearly separable, using a kernel. The use of kernels has recently received significant attention in learning systems. The primary explanation for this is that kernels enable data to be mapped into a high-dimensional space to increase linear machines' computing power [1–3]. Therefore, it is a way of extending linear hypotheses; this step can be performed implicitly for non-linear ones. Continuous research shows how various validated architectures in machine learning field, such as neural networks, decision trees, or k-nearest neighbor, may outperform support vector machine. They construct models that are complex enough to solve real-world applications that remain comfortable enough to be mathematically tested. One interpretation of the efficacy of the kernel-based approach is the acceptance of the implicit relationship in the data by the kernel function and converting it to explicit, thus resulting in patterns which are easier to recognize [4]. Thus, each kernel function extracts information about a particular kind and a partial explanation or view of the data is thus generated from a specified dataset [5].

DOI: 10.1201/9781003303053-12

Although SVM enjoys high popularity in classifying linear separable data, our optimization objective is to maximize the margin. It is commonly used for classifying non-linearly separable data using a kernel. This paper illustrates how mixing two or more kernels will increase the accuracy of any dataset's estimation of an outcome. Therefore, it is essential to define the high-quality kernel function to fully achieve an SVM with high precision classification. The proposed function of the kernel blends the polynomial and Gaussian kernels and is analyzed to test their dominance over these individual kernels over a simple dataset that uses logical or a function from NumPy [6].

To achieve the computational power, Shogun (Python interface), a machine learning toolbox implemented in C++ that focuses primarily on large-scale kernel methods and, in particular, on SVM, is used. It comes up with several different support vector machine implementations with a similar SVM object interface, all using the same underlying powerful kernel implementation [7]. Since Shogun is still in the development phase, the multiple kernel learning classification – MKLClassification algorithm – only supports SVMLight. Here, SVMLight [8] is an implementation of support vector machine in the programming language C and a solver for kernel SVMs. To solve a series of optimization issues by reducing the solution, the algorithm uses a local search approach to train transductive large-scale SVMs [9].

The outcome of this experiment mostly relies on the type of kernel chosen for the combination, and it changes according to the dataset. The proposed multi-kernel in this chapter proves that multi-kernel SVM performs much better than single-kernel SVMs.

In Section 10.2, the mathematics behind the formulation of the vector machine is briefly discussed based on two cases in the mapping of the dataset. The main purpose of this paper is to discuss in the second case, i.e., Section 10.2, a non-linearly separable case that uses the kernel to solve the problem that is discussed in the subsections of 10.2. Section 10.3 consists of the analysis of the experiment on a sample dataset and their result, and Section 10.4 is a brief conclusion based on the results obtained.

10.2 SUPPORT VECTOR MACHINE STATISTICS

SVMs have initially been proposed for solving binary classification problems [10]. Thusly, the intuition behind SVMs will be clarified on a binary class dataset. Binary sorting is the easiest of all classification operations. Let us consider a given training dataset D of N samples with x_i and y_i as input vector and a binary classification label:

$$D = \{(x_i, y_i) \mid x_i \in \{1, -1\}\}_{i=1}^{n} \tag{10.1}$$

A function f that accepts x input sets and defines the output $y = f(x)$ and can be used as a decision limit or a hyperplane (w, b) dividing the input data into two regions [11]:

$$f(x) = \langle w, x \rangle + b \tag{10.2}$$

The performance of the classifier can be measured as:

$$error(f(x), y) = \begin{cases} 0 & \text{if } f(x) = y \\ 1 & \text{otherwise} \end{cases} \tag{10.3}$$

In linear classifier, there are two cases based on their separability. For $x \rightarrow f(x, a)$ mapping, dataset is separable, and non-separable in case of non-feasible perfect mapping.

10.2.1 Separable case

Consider a classification set of binary separable data that can be segregated, as shown in Figure 10.1a. The cross-marks are data points that belong to the target value $y_i = +1$ and can be denoted by S_+ and circle marks belong to $y_i = -1$ and can be denoted by S_-. The mapping that can separate the S_+ and S_- is as follows:

$$f(x, y) = \sin(\bar{w}, \bar{x} + b) \tag{10.4}$$

where $\bar{w}$ is the weight vector and b is the bias. The functional instance of the hyperplane would change depending on $\bar{w}$ and b. There can be multiple

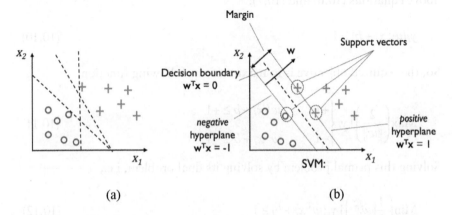

Figure 10.1 (a) Multiple separating hyperplanes; (b) the optimal separating hyperplane.

values of $\{w,b\}$ that satisfy the equation of decision boundary. By maximizing the margin between the two support vectors, SVM finds the optimal hyperplane for S_+ and S_- as shown in Figure 10.1b.

Given such a mapping, the hyperplane (decision boundary) can be defined as [12]:

$$\vec{w}^{T}\vec{x}+b = 0 \tag{10.5}$$

Separating the positive and negative data points such that

$$\vec{w}^{T}\vec{x}+b \geq 1 \quad \text{for } y_i = +1 \tag{10.6}$$

$$\vec{w}^{T}\vec{x}+b \leq -1 \quad \text{for } y_i = -1 \tag{10.7}$$

the margin gap can be determined by subtracting the two equations (10.6) and (10.7) from the scalar view, i.e.,

$$w^{T}(x_{pos} - x_{neg}) = 2 \tag{10.8}$$

Here, $(x_{pos} - x_{neg})$ is the margin that is to be maximized. To get rid of the w^{T} magnitude, the w vector norm $(\|w\|)$ is broken into both sides [13]. Equation (10.8) now becomes:

$$\frac{w^{T}}{\|w\|}(x_{pos} - x_{neg}) = \frac{2}{\|w\|} \tag{10.9}$$

This acquired margin $\left(\dfrac{2}{\|w\|}\right)$ is to be maximized to get an optimal hyperplane in such a way that it should follow the constraint acquired from the above equations (10.6) and (10.7):

$$y_i(w^{T}x_i + b) \geq 1 \tag{10.10}$$

So, the required objective is to maximize the following function:

$$\text{Max}\left(\frac{2}{\|w\|}\right) | y_i \begin{cases} +1 & w^{T}x_i + b \geq +1 \\ -1 & w^{T}x_i + b \leq -1 \end{cases} \tag{10.11}$$

Solving this primal problem by solving its dual problem, i.e.,

$$\text{Min}\left(\frac{1}{2}\|w\|^{2}\right) | y_i(w^{T}x_i + b) \geq 1 \tag{10.12}$$

next, the constraints on the Langrage multipliers will be replaced by restrictions that will be much easier to work with and second, the training data in this first issue formulation will only take place in the form of products of data among the vectors. This will enable us to reason out the non-separable case method. The Lagrange is as follows:

$$l = \frac{1}{2}(W,W) - \sum_{i=1}^{m} \alpha_i(y_i(wx_i + b) - 1)$$ (10.13)

Thus, the Lagrange multipliers are $\alpha_i, i = 1,2,...,m$. If we set

$$\frac{\partial l}{\partial w} = 0, \frac{\partial l}{\partial b} = 0,$$ (10.14)

$$w = \sum_{i=1}^{m} \alpha_i x_i y_i = 0$$ (10.15)

Evaluating (10.13) back to obtain weight vector that is maximized w.r.t α_i,

$$w(\alpha) = \sum_{i=1}^{m} \alpha_i - \sum_{i,j=1}^{m} \alpha_i \alpha_j y_i y_j (x_i, x_j)$$ (10.16)

subject to:

$$\sum_{i=1}^{m} \alpha_i y_i = 0 \ \forall \ \alpha_i \geq 0$$ (10.17)

On solving this optimization problem, the model can successfully predict the unseen data.

10.2.2 Non-separable case

To operate on $x \rightarrow f(x,a)$ mapping, the classic SVMs are formulated, but they don't seem to work on a non-linearly separable dataset. An extension ("slack" variable $\varsigma \geq 0$) to the above formula is then added to treat non-separable data by controlling the misclassification against the minimization of $\|w\|^2$ [14]:

$$w \cdot x_i - b \geq +1 - \varsigma \text{ for } y_i = +1$$ (10.18)

$$w \cdot x_i - b \geq -1 + \varsigma \text{ for } y_i = -1$$ (10.19)

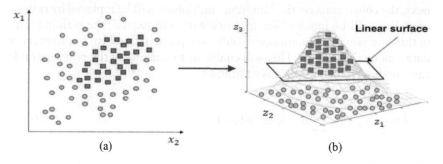

Figure 10.2 (a) Non-linearly separable feature space; (b) projected linearly separable feature space.

10.2.2.1 Kernel method

Kernel functions are used with non-linearly separable datasets for the creation of original feature's non-linear combination that can be projected to a higher-dimensional space using the φ projection algorithm, where they become linearly separable as shown in Figure 10.2.

$$\phi(x_1, x_2) = (z_1, z_2, z_3) = (x_1, x_2, x_1^2 + x_2^2) \tag{10.20}$$

Figure 10.2a is a non-linearly separable two-dimensional data mapping that is converted into a linearly separable one with the help of some combination functions as shown in Figure 10.2b.

In the above discussion, it is proved that if we have a non-linearly separable dataset, one could create a function (a kernel function) that could generate more dimensions that are based on those previous features. So, there is only one major element to kernels; that is, they are obtained using the inner products. Let's say the X-space feature set $(x_1, x_2, \ldots x_m)$, m being the dimension of an example in a feature set, can be written in Z-space feature set $(z_1, z_2, \ldots z_k)$ k as a new dimension of the example in that feature set as demonstrated in the above equation (10.20). Earlier in the mapping equation $f(x, y) = \sin(\bar{w}.\bar{x} + b)$, $(\bar{w}, \bar{x})$ is the inner product, which states that the x features can be replaced by z features on the basis that an inner product returns a scalar value. This depicts that the dimension of the space is of no significance for the output y. Hence, the weight vector that's stated in equation (10.16) is itself a scalar value with vector dot product:

$$w(\alpha) = \sum_{i=1}^{m} \alpha_i - \sum_{i,j=1}^{m} \alpha_i \alpha_j y_i y_j (x_i.x_j) \tag{10.21}$$

Kernel methods work by converting the data items of feature space and searching linear relations in such a space by finding a functional relationship between them [15]. This modularization is implicitly specified by defining an internal product through a positive semi-definite kernel function for the feature space [16]:

$$K(x,z) = \langle \Phi(X) \cdot \Phi(Z) \rangle \qquad (10.22)$$

10.2.2.2 Multiple kernel learning

Traditional single-kernel SVMs are commonly used in many areas, but in some important decision-making applications, such as load forecasting in power systems or fault analysis in power systems, they are not that detailed enough. Multiple kernel learning consists of creating the appropriate kernel function for each data type and using these kernel functions in separate kernel-based models to achieve an integrative interpretation of the varying data type and a better estimation of the data type [17]. MK functions have two primary forms: linear and non-linear. The linear MK function's basic form is given by [18]:

$$K_{MKL}(x,z) = \sum_{h=1}^{m} \beta_h K_h(x,z) \qquad (10.23)$$

where K_{MKL} is the combined kernel, m is the number of sub-kernels, K_h is an individual kernel, and $\beta_h > 0, \sum_{h=1}^{m} \beta_h = 1$.

Non-linear multiple kernel function basic form is given by:

$$K_{MKL}(x,z) = \beta_1 K_1(x,z) \times (\beta_2 K_2(x,z) + \beta_3 K_3(x,z))$$
$$+ \beta_4 K_4(x,z) + \qquad (10.24)$$

Kernels used by the SVM are reported to be divided into two classes: local and global kernels. Only those near the test points significantly impact kernel values in the local kernel, whereas in the global kernel, points far away from the test point have the most impact [19]. The RBF (Gaussian) and polynomial kernel functions are two classic examples of local and global kernels, respectively. Combined with local and global kernel functions with different deduction capabilities, the MK function will facilitate strengths and bypass shortcomings and achieve a better output SVM prediction model [20]. Therefore, according to (10.23), it is possible to construct a type of SVM modeling system based on a kernel mixture as follows:

$$K_{Mix}(x,z) = \beta K_{RBF}(x,z) + (1 - \beta) K_{Poly}(x,z) \qquad (10.25)$$

where $K_{RBF}(x,z)$ and $K_{Poly}(x,z)$ denote the function of the RBF kernel and polynomial kernel with their respective formulas shown in equations

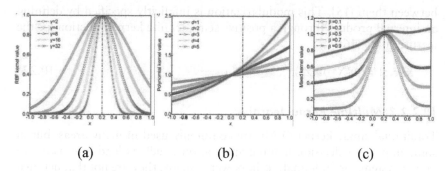

Figure 10.3 (a) RBF kernel function curve, (b) polynomial kernel function curve, and (c) function curve for the proposed mixed kernel.

(10.26) and (10.27), $K_{\mathrm{Mix}}(x,z)$ is the composite function of the kernel, and $\beta(0 \le \beta \le 1)$ is the mixed coefficient used to change the weights of the two kernel function forms.

$$K_{\mathrm{RBF}}(x,z) = \exp\left(\frac{\|x-z\|^2}{2\sigma^2}\right) \text{ where } \frac{1}{2\sigma^2} = \gamma \tag{10.26}$$

$$K_{\mathrm{Poly}}(x,z) = (x^T z + c)^d \tag{10.27}$$

Figure 10.3a shows the change in the curve of RBF kernel function with the change in its γ value with respect to feature value x. Similarly, Figure 10.3b shows the change in the polynomial function curve with the change in the value of d w.r.t feature value x. Figure 10.3c shows the superposition of graphs (10.3a) and (10.3b), which depicts the combination of both the kernels which can also be plotted mathematically with the help of equation (10.25), where β is the kernel weight.

10.3 EXPERIMENT AND RESULTS

For examination, on a basic dataset that has the state of an XOR entryway utilizing the logical or function from NumPy, we assess the proposed multi-kernel against the individual portions of the mixed kernel. Shogun's multiple kernel learning is executed by adjusting the source code of LibSVM (a library for support vector machine). Shogun's MKLClassification method supports only SVMLight (an SVM solver that takes care of the streamlining issues by lower-bounding the arrangement utilizing a type of local search [21,22]) as an input to develop the combination of RBF and polynomial part. We performed tenfold cross-validation of the training dataset to decide the ideal characterization boundary β and if the best boundaries are utilized in testing. The comparing boundaries are set here as $\gamma=15$, $c=1$, $d=1$, and the best agent estimation of β will be set by the Shogun MKLClassification algorithm itself.

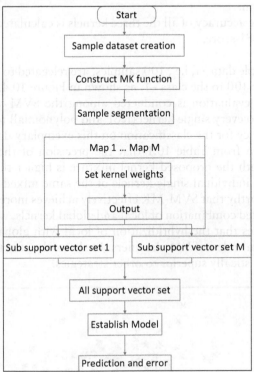

Input: The sample dataset (non-linearly separable) with 200 input samples and 1 feature set with the corresponding sample contains the target as 1 or –1.

Experiment: Prediction of the target values corresponding to the sample data correctly and calculating the accuracy.

Step 1: Forming a 200-sample dataset by randomizing the feature values with NumPy random function and assigning random 100 samples with +1 and rest with –1.

Step 2: Creating an instance of the kernels (RBF, polynomial) for individual application of these kernels using Shogun toolbox.

Step 3: Sample segmentation into 10 segments for cross-validation (Map 1… Map 10).

Step 4: Shogun toolbox is used to create an instance of the combined kernel by appending the individual kernels to the list.

Step 5: An instance of multiple kernel learning classification function is made by passing the SVM solver (SVMLight) as the parameter (kernel weights) to solve the complex calculations involved in the process.

Step 6: The combined kernel is now set to the SVM-MK classification model by using (instance of MKL Classification.set_ kernel) function.

Step 7: The accuracy of all the three kernels is calculated with the help of the F1-score.

Half of the sample dataset, i.e., 100 samples, are relegated to the class mark 1, and the other 100 to the class –1, as shown in Figure 10.4.

An accuracy evaluation is conducted among the SVM and SVM-MK with reference to every single kernel (RBF and polynomial) to illustrate the SVM-MK potency for the classification on this exemplary dataset.

It can be seen from Table 10.1 that the precision of the classification of SVM-MK with the proposed kernel mixture is higher than that of the SVM type with individual single kernels of the same mixed mixture pool. It is also noteworthy that SVM-MK effectively achieves more robust results with the suggested combination of local and global kernels, which additionally demonstrates that the hybridization of local with global kernel functions plays a vital role in SVM classifiers. In short, the effectiveness of our technique is holistically superior to other strategies.

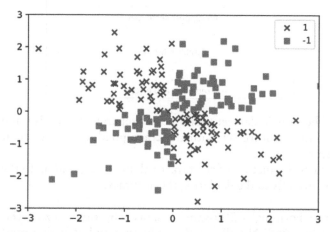

Figure 10.4 Non-linear separable dataset mapping used for the experiment.

Table 10.1 Comparison of SVM-RBF, poly, and MK classification accuracy on the applied dataset

Classifier	SVM-RBF	SVM-Poly	SVM-MK
Accuracy (%)	71.67	56.67	82.34

10.4 CONCLUSIONS

In this chapter, we proposed a novel architecture for support vector machine with a kernel mix (SVM-MK) for the efficient process of classification. But it may also apply to regression models. The combined global and local kernels' output will open up many possibilities for the same combination. Besides, the experimental findings on this artificial dataset show that the proposed classifier can be applied to any form of a dataset consisting of both local and global segmentation, SVM-MK, which, with many state-of-the-art approaches, can achieve a greater classification precision compared to any standard single-kernel SVM, as well as comparable performance.

REFERENCES

1. D. P. Mishra and P. Ray, "Fault detection, location and classification of a transmission line," *Neural Computing and Applications*, Vol. 30, no. 5, pp. 1377–1424, 2018.
2. V. Vapnik, "The support vector method of function estimation," in Suykens, J.A.K., and Vandewalle, J. (eds), Nonlinear Modeling. Springer, Boston, MA, pp. 55–85, 1998. https://doi.org/10.1007/978-1-4615-5703-6_3
3. N. Cristianini and J. Shawe-Taylor, *An Introduction to Support Vector Machines and Other Kernel-Based Learning Methods*, Cambridge University Press, London, 2000.
4. G. R. G. Lanckriet, T. De Bie, N. Cristianini, M. I. Jordan and W. S. Noble, "A statistical framework for genomic data fusion," *Bioinformatics*, Vol. 20, no. 16, pp. 2626–2635, 2004.
5. A. Smolinska, L. Blanchet, L. Coulier, K. A. M. Ampt, T. Luider, R. Q. Hintzen, S. S. Wijmenga and L. M. C. Buydens, "Interpretation and visualization of non-linear data fusion in kernel space: Study on metabolomic characterization of progression of multiple sclerosis," *PLoS One*, Vol. 7, no. 6, p. e38163, 2012.
6. A. Ashraf, E. Zanaty and G. Said, "Improving the classification accuracy using Support Vector Machines (SVMs) with new kernel," *Journal of Global Research in Computer Sciences*, Vol. 4, pp. 1–7, 2013.
7. P. Ray and D. P. Mishra, "Support vector machine based fault classification and location of a long transmission line," *Engineering Science and Technology, an International Journal*, Vol. 19, pp. 1368–1380, 2016.
8. "SVM-Light: Support vector machine" http://svmlight.joachims.org/.
9. R. Kilany, R. Ammar and S. Rajasekaran, "A novel algorithm for technical articles classification based on gene selection," in *2012 IEEE Symposium on Computers and Communications (ISCC)*, pp. 000234–000238, 2012. doi: 10.1109/ISCC.2012.6249300.
10. P. Ray, D. P. Mishra and D. D. Panda, "Hybrid technique for fault location of a distribution line," in *2015 Annual IEEE India Conference (INDICON)*, New Delhi, pp. 1–6, 2015, Doi: 10.1109/INDICON.2015.7443134.

11. L. Dioşan, A. Rogozan and J.-P. Pécuchet, "Evolutionary optimisation of kernel and hyper-parameters for SVM," In Le Thi, H.A., Bouvry, P., and Pham Dinh, T. (eds), *Modelling, Computation and Optimization in Information Systems and Management Sciences. MCO 2008. Communications in Computer and Information Science*, Vol. 14. Berlin, Heidelberg: Springer, pp. 107–116, 2008. https://doi.org/10.1007/978-3-540-87477-5_12

12. M. Hofmann, "Support vector machines — Kernels and the kernel trick an elaboration for the Hauptseminar "Reading Club: Support Vector Machines"", Computer Science, pp. 1–16 2006.

13. H. Abdi and W. Lynne, "Normalizing data," *Encyclopedia of Research Design* Vol. 1, pp. 935–936 2010.

14. A. Zanaty and S. Aljahdali, "Improving the accuracy of support vector machines," in *Proceedings of the ISCA 23rd International Conference on Computers and Their Applications, CATA* 2008, 9–11 April 2008, Cancun, Mexico.

15. G. R. G. Lanckriet, M. Deng, N. Cristianini, M. I. Jordan and W. S. Noble, "Kernel-based data fusion and its application to protein function prediction in yeast," in *Pacific Symposium on Biocomputing*, Hawaii, pp. 300–311, 2004.

16. M. Genton, "Classes of kernels for machine learning: A statistics perspective," *Journal of Machine Learning Research*, Vol. 2, pp. 299–312, 2001.

17. S. Amari and S. Wu, "Improving support vector machine classifiers by modifying kernel functions," *Neural Network*, Vol. 12, no. 6, pp. 783–789, 1999.

18. P. Ray and D. P. Mishra, "Signal processing technique based fault location of a distribution line," in *2015 IEEE 2nd International Conference on Recent Trends in Information Systems (ReTIS)*, Kolkata, pp. 440–445, 2015.

19. D. Tian, X. Zhao and Z. Shi, "Support vector machine with mixture of kernels for image classification," In Shi, Z., Leake, D., and Vadera, S. (eds), *Intelligent Information Processing VI*, Berlin, Heidelberg: Springer, pp. 68–76, 2012.

20. Q. Ren, M. Li, L. Song and H. Liu, "An optimized combination prediction model for concrete dam deformation considering quantitative evaluation and hysteresis correction," *Advanced Engineering Informatics*, Vol. 46, p. 101154, 2020.

21. B. Léon, C. Olivier, D. Dennis and W. Jason, "Support vector machine solvers," *Large-Scale Kernel Machines*, pp. 1–27, 2007.

22. C.-J. Hsieh, S. Si and I. S. Dhillon, "A divide-and-conquer solver for kernel support vector machines," in *Proceedings of the 31st International Conference on International Conference on Machine Learning*, Vol. 32, pp. 566–574, 2014.

Chapter 11

An introduction to parallelisable string-based SP-languages

N. Mohana and Kalyani Desikan
Vellore Institute of Technology

V. Rajkumar Dare
Madras Christian College

CONTENTS

11.1 INTRODUCTION

A sequential task is computationally modelled as a concatenation wherein a system finishes one task before starting the next one. Parallel computing is an important paradigm in computer architecture. Parallelism uses multiple resources to complete sequential tasks simultaneously. Parallel computers are classified based on the stage at which parallelism is promoted by the hardware.

In general, not all tasks can be parallelised. There are some tasks that should be completed sequentially. Formal language helps to represent tasks in terms of strings, where each task can be represented as an alphabet and sequence of tasks can be considered as a string. Parallel operator is used to indicate that some sequential tasks are happening in parallel.

Automaton [1–4] is one of the tools (machine) used to do computations, but it normally works only on sequential tasks. Pomsets are attained from singletons by sequential and parallel products and also by the trace product. Parallelism on labelled partially ordered sets or pomsets (partially ordered multisets) were put forth by Lodaya and Weil [5–8]. Havel [9]

DOI: 10.1201/9781003303053-13

first introduced branching automata, and their logical methods have been derived by Beden in [10]. N-free pomsets that have bounded width were introduced by Lodaya and Weil [6], and their properties have been extended to infinite pomsets in [11]. Recently, algebraic properties have been proved in [12] for the series-parallel posets. SP-languages can be widely used in many areas such as parallel computation, fork-join programming, series-parallel circuits, and series-parallel graphs. Fork-join model has been used in queueing theory [13] to identify the maximum time of the parallel queue.

11.2 PARALLELISABLE STRING-BASED SP-LANGUAGES

Let Σ be a finite alphabet and Σ^* be the collection of all words (strings) over Σ. Let Σ_p^* be the set of all parallel-series strings, $\Sigma s^\oplus$ be the collection of all series-parallel strings, and $\Sigma^\oplus$ be the set of all parallel strings over Σ.

In general,

$$\Sigma^\oplus = \{y_1\|y_2\|y_3\|...\|y_m : y_i \in \Sigma \cup \{\epsilon\}, 1 \leq i \leq m\}$$

$$\Sigma_p^* = \{y_1y_2y_3...y_m : y_i \in \Sigma^\oplus, 1 \leq i \leq m\}.$$

$$\Sigma s^\oplus = \{y_1\|y_2\|y_3\|...\|y_m : yi \in \Sigma^*, 1 \leq i \leq m\}.$$

Definition 11.1

Consider SP $(\Sigma) = \Sigma_p^* \cup \Sigma s\oplus$. Here, SP$(\Sigma)$ is the set of all parallelisable string-based SP-strings over Σ. $L \subset SP (\Sigma)$ is defined as a parallelisable string-based SP-language.

Definition 11.2

Length l and depth d of a parallelisable string-based SP-language are defined as follows.

Let $z, w \in SP (\Sigma)$, where $w = u\|v$, $z = x \cdot y$ with $u, v \in \Sigma^*$ and $x, y \in \Sigma\oplus$.

$$l(z) = l(x) + l(y) \text{ and } l(w) = \max(l(u), l(v))$$

$$d(z) = \max(d(x), d(y)) \text{ and } d(w) = d(u) + d(v)$$

11.3 PARALLEL REGULAR EXPRESSION

Definition 11.3

If r and s are REs, then

- The parallelism of r and s, $r\|s$ is a parallel expression. It is defined as the collection of all strings from r and s arranged in parallel, which is represented in Figure 11.1.

- The parallel closure of r, $r^{\oplus}$ is a parallel regular expression. It is defined as at least one occurrence of r followed in parallel by any number of occurrences of r, which is represented in Figure 11.2.

Definition 11.4

Let r and s be REs or parallel regular expressions and $L(r)$ and $L(s)$ be the corresponding regular languages or parallel regular languages, then $L(r^{\oplus}) = (L(r))^{\oplus}$ and $L(r\|s) = L(r)\|L(s)$.

Example 11.1

If $r = \{a\}$ and $s = \{b\}$, then $L(s\|r) = b\|a$ and $L(r\|s) = a\|b$.

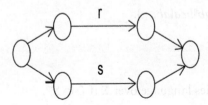

Figure 11.1 Automaton representation for $r\|s$.

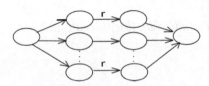

Figure 11.2 Automaton representation of $r^{\oplus}$.

Example 11.2

If $r = a+b$ and $s = b+a$, then $L(s) = L(b) \cup L(a)$ and $L(r) = L(a) \cup L(b)$. Now $L(s\|r) = L(s)\|L(r) = \{b\|a, b\|aa, ab\|a, ab\|aa\}$.

$L(r\|s) = L(r)\|L(s) = \{a\|b, a\|ab, aa\|b, aa\|ab\}$.

Definition 11.5

$L^{\oplus}$ is a parallel language over Σ if $L \subset \Sigma^{\oplus}$.

Example 11.3

Let $L = \{a\|b, b\|c\}$, then

$L^{\oplus} = \{(a\|b)\|(a\|b), (b\|c)\|(b\|c), (a\|b)\|(a\|b)\|(b\|c),...\}$ which is same as $L^{\oplus} = \{a\|b\|a\|b, b\|c\|b\|c, a\|b\|a\|b\|b\|c,...\}$.

Definition 11.6

$Ls^{\oplus}$ is a series-parallel language over Σ if $L \subset \Sigma^*$.

Example 11.4

Let $L = \{ab, bca\}$, then

$Ls^{\oplus} = \{ab\|ab, ab\|bca\|ab,...\}$.

Definition 11.7

Lp^* is a parallel-series language over Σ if $L \subset \Sigma^{\oplus}$.

Example 11.5

Let $L = \{a\|b, b\|c\}$, then

$Lp^* = \{(a\|b)(b\|c), (a\|b)(a\|b), (b\|c)(b\|c)(a\|b),...\}$.

Definition 11.8

A branching automaton [6] over Σ is $B = (Q, \Sigma, \delta_{seq}, \delta_{fork}, \delta_{join}, S, E)$, where Q is the collection of finite states. $\delta_{seq} \subseteq Q \times \Sigma \times Q$, $\delta_{fork} \subseteq Q \times M_n(Q)$, and $\delta_{join} \subseteq M_n(Q) \times Q$ are the collection of sequential, fork, and join transitions. Here, $M_n(Q)$ (non-singleton and non-empty) stands for multisets over Q

of size at least 2. S and E are the finite collection of initial and final states, respectively.

If $w = a_1 a_2 a_3 \ldots a_n$ is a string, then $\delta_{\text{seq}}(q_0, w) = q_n$.

If $w = a_1 \| a_2 \| a_3 \| \ldots \| a_n$ is a parallel string, then

$$\delta_{\text{join}} \delta_{\text{seq}} \delta_{\text{fork}}(q_0, a_1 \| a_2 \| a_3 \| \ldots \| a_n) = \delta_{\text{join}} \delta_{\text{seq}}(\{q_{01}, q_{02}, q_{03}, \ldots, q_{0n}\}, a_1 \| a_2 \| a_3 \| \ldots \| a_n)$$

$$= \delta_{\text{join}} \delta_{\text{seq}}(\{(q_{01}, a_1), (q_{02}, a_2), \ldots, (q_{0n}, a_n)\})$$

$$= \delta_{\text{join}}(\{q_{n1}, q_{n2}, q_{n3}, \ldots, q_{nn}\})$$

$$= q_n$$

where $\{q_{01}, q_{02}, q_{03}, \ldots, q_{0n}\}, \{q_{n1}, q_{n2}, q_{n3}, \ldots, q_{nn}\} \in M_n(Q)$.

An input w is recognised by branching automaton $\mathbf{B}$ if w starts at the start state and ends in an end state of $\mathbf{B}$. A language L is said to be regular if it is the collection of strings recognised by a finite branching automaton $\mathbf{B}$. We state that $\mathbf{B}$ accepts L and denote $L = L(\mathbf{B})$.

Example 11.6

Consider $B = (Q, \Sigma, \delta_{\text{seq}}, \delta_{\text{fork}}, \delta_{\text{join}}, S, E)$ a branching automaton where $Q = \{q_0, q_1, q_2, \{q_{21}, q_{22}\}, \{q_{41}, q_{42}\}, q_4\}$, $\Sigma = \{a, b\}$, $S = \{q_0\}$, $E = \{q_4\}$,

$\delta_{\text{seq}} = \{(q_0, a, q_1), (q_1, b, q_2), (q_{21}, a, q_{41}), (q_{22}, a, q_{42})\}$, $\delta_{\text{fork}} = \{(q_2, \{q_{21}, q_{22}\})\}$ and $\delta_{\text{join}} = \{(\{q_{41}, q_{42}\}, q_4)\}$.

The language $L = \{(ab(aa))^m : m > 0\}$ is recognised by branching automaton $\mathbf{B}$ in Figure 11.3.

Definition 11.9

$L \subseteq \text{SP}(\Sigma)$ a parallelisable string-based SP-language is regular if a branching automaton $\mathbf{B}$ recognises it.

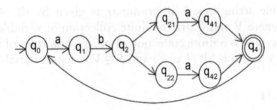

Figure 11.3 Branching automaton for a language $L = \{(ab(a\|a))^m : m > 0\}$.

We defined that if $x = x_1 x_2 x_3 \ldots x_m$, where $x_i \in \Sigma^{\oplus}$, $1 \leq i \leq m$, then each x_i has transitions as given for the branching automaton. And concatenation (series) of $x_1, x_2, x_3, \ldots, x_m$ is achieved by connecting an end state of x_i to an initial state of x_j for $1 \leq i \leq m$ and $j = i+1$. If $x = x_1 \| x_2 \| x_3 \| \ldots \| x_m$, where $x_i \in \Sigma^*$, $1 \leq i \leq m$, then each x_i has transition $\delta_{seq}(q_i, x_i) = q'_i$ for $1 \leq i \leq m$. Now parallelism of $x_1, x_2, x_3, \ldots, x_m$ involves the fork and join transitions. Hence, a parallelisable string-based SP-language is recognised by a branching automaton.

11.4 EQUIVALENCE OF PARALLEL REGULAR EXPRESSION AND BRANCHING AUTOMATON

There is equivalence between finite automata and regular expressions in their descriptive power. Any regular expression that describes a language can be converted into a FA that recognises the language and vice versa. For parallelism and parallel closure, we have used finite branching automaton instead of finite automaton. Equivalence between regular expression and finite automaton is given in [1].

Theorem 11.1

Let s be a parallel regular expression. Then there exists a branching automaton that accepts $L(s)$ if $L(s)$ is of bounded depth. Consequently, $L(s)$ is a parallelisable string-based SP-regular language.

Proof. We begin with an automaton that accepts the language for simple regular expressions φ, s, and $a \in \Sigma$. The automaton representation has been given in [1].

Also, we have given the automaton representation of $r \| s$, $r+s$, $r^{\oplus}$, and r^* in [1]. Figures 11.1 and 11.2 show the automaton representation of $r\,s$ and $r^{\oplus}$ of bounded depth. It states that every parallel regular expression has an equivalent branching automaton that accepts a language of bounded depth. Hence, $L(s)$ is a parallelisable string-based SP-regular language.

11.5 PARALLELISABLE STRING-BASED SP-GRAMMAR

Definition 11.10

A parallelisable string-based SP-grammar is given by the 4-tuple $G_{sp} = (V, T, P, S)$, where V and T are the finite collection of variables and terminals, respectively. P is a finite collection of production rules of the form $\beta \rightarrow \alpha$, where $\beta \in V$ and $\alpha \in SP\,(T \cup V)$ and $S \in V$ is a start variable.

Definition 11.11

A grammar $G_{sp} = (V, T, P, S)$ is a right-linear grammar if P contains $V_i \rightarrow x \| V_j | x V_j$ or $V_i \rightarrow x$ with $V_i, V_j \in V$ and $x \in T$.

Example 11.7

Let G_{sp} be a parallelisable string-based SP-grammar with $T = \{a, b\}$, $V = \{S, A, B\}$, S being a start variable, and $P = \{S{\rightarrow}bA, B{\rightarrow}bA, A{\rightarrow}a\|B, A{\rightarrow}a\}$.

The sequences of some derivations are given below.

$S \Rightarrow bA \Rightarrow ba$

$S \Rightarrow bA \Rightarrow ba\|B \Rightarrow ba\|bA \Rightarrow ba\|ba$

$S \Rightarrow bA \Rightarrow ba\|B \Rightarrow ba\|bA \Rightarrow ba\|ba\|B \Rightarrow ba\|ba\|bA \Rightarrow ba\|ba\|ba$

$L(G_{sp}) = \{(ba)^{m\oplus} : m > 0\}$ is the language generated by G_{sp}, and the corresponding parallel regular expression is $r = (ba)^{\oplus}$.

Definition 11.12

A grammar $G_{sp} = (V, T, P, S)$ is a left-linear grammar if P contains

$$V_i \rightarrow V_j\|x|V_jx \text{ or } V_i \rightarrow x$$

with $V_i, V_j \in V$ and $x \in T$.

Example 11.8

Let $G_{sp} = (V, T, P, S)$ be a parallelisable string-based SP-grammar where $V = \{S, A, B\}$, $T = \{a, b\}$, S is a start variable, and $P = \{S \rightarrow A\|a, A \rightarrow Ba, B \rightarrow A\|a, A \rightarrow a\}$. The sequences

$S \Rightarrow A\|a \Rightarrow a\|a$

$S \Rightarrow A\|a \Rightarrow Ba\|a \Rightarrow A\|a{\cdot}a\|a \Rightarrow a\|a{\cdot}a\|a$

$S \Rightarrow A\|a \Rightarrow Ba\|a \Rightarrow A\|a{\cdot}a\|a \Rightarrow Ba\|a{\cdot}a\|a \Rightarrow A\|a{\cdot}a\|a{\cdot}a\|a \Rightarrow a\|a{\cdot}a\|a{\cdot}a\|a$

are some derivations of G_{sp}.

$L(G_{sp}) = \{(a\|a)^m : m > 0\}$ is the language generated by G_{sp}, and the corresponding parallel regular expression is $r = (a\|a)^+$.

11.6 PARALLELISABLE STRING-BASED SP-PARALLEL GRAMMAR

Definition 11.13

A parallelisable string-based SP-parallel grammar is given by the 4-tuple $G_{sp} = (V, T, P, S)$, where $T \subseteq \Sigma \cup \Sigma^{\oplus}$, V is a finite set of variables, $S \in V$ is a start variable, and P is a finite collection of productions $\beta \rightarrow \alpha$ with $\beta \in V$ and $\alpha \in (T \cup V)^*$.

Definition 11.14

A grammar G_{sp} is defined to be parallelisable string-based SP-parallel left-linear if the productions are of the form $V_i \rightarrow V_j x$ (or) $V_i \rightarrow x$ with $x \in T$ and $V_i, V_j \in V$.

Definition 11.15

A parallelisable string-based SP-left linear grammar is defined to be parallelisable string-based SP-parallel left linear grammar if its production can be written as follows:

- Every left-linear production $V_i \rightarrow V_j \| x$ and $V_j \rightarrow y$ is equivalent to a parallel left-linear production $V_i \rightarrow y \| x$.
- Every left-linear production $V_i \rightarrow V_j \| x$ and $V_j \rightarrow V_{jy}$ is equivalent to a parallel left-linear production $V_i \rightarrow V_j(y \| x)$.

where $V_i, V_j \in V$ for some i, j and $x, y \in T$

Example 11.9

Let $G_{sp} = (V, T, P, S)$ be a parallelisable string-based SP-grammar where $T = \{b, a\}$, $V = \{S\}$, S is a start variable, and $P = \{S \rightarrow S(b\|a), S \rightarrow (b\|a)\}$.
    The sequences of some derivations are as follows.

$S \Rightarrow (b\|a)$

$S \Rightarrow S(b\|a) \Rightarrow (b\|a)(b\|a)$

$S \Rightarrow S(b\|a) \Rightarrow S(b\|a)(b\|a) \Rightarrow (b\|a)(b\|a)(b\|a)$

$L(G_{sp}) = \{(b\|a)^m : m > 0\}$ is the language generated by G_{sp}, and the corresponding parallel regular expression is $r = (b\|a)^+$.

Example 11.10

Let $G_{sp} = (V, T, P, S)$ be a parallelisable string-based SP-grammar, where $T = \{b\}$, $V = \{S, A, B\}$, S is a start variable, and P is the collection of productions $S \rightarrow Ab$, $B \rightarrow Ab$, $A \rightarrow B\|b$, $A \rightarrow b$. An equivalent parallel left-linear SP-grammar G_{sp}, where $T = \{b, b\|b\}$, $V = \{S\}$, S is a start variable, and $P = \{S \rightarrow Sb, S \rightarrow S(b\|b), S \rightarrow b\}$.
    The sequences

$S \Rightarrow Sb \Rightarrow bb$

$S \Rightarrow Sb \Rightarrow S.b \| b.b \Rightarrow bb \| bb$

$S \Rightarrow Sb \Rightarrow S.b \| b.b \Rightarrow S.b \| b.b \| b.b \Rightarrow bb \| bb \| bb$

are some derivations of G_{sp}.

$L(G_{sp}) = (bb)^{m \oplus} : m > 0$ is the language produced by G_{sp}, and its corresponding parallel regular expression is $(bb)^{\oplus}$.

Definition 11.16

A grammar G_{sp} is defined to be parallelisable string-based SP-parallel right-linear if the collection of all productions are of the form $V_i \to x V_j$ (or) $V_j \to x$ with $x \in T$ and $V_i, V_j \in V$.

Definition 11.17

A parallelisable string-based SP-right linear grammar is defined to be parallelisable string-based SP-parallel right linear grammar if its production can be written as follows:

- Every right-linear production $V_i \to x \| V_j$ and $V_j \to y$ is equivalent to a parallel right-linear production $V_i \to x \| y$;
- Every right-linear production $V_i \to x \| V_j$ and $V_j \to y V_j$ is equivalent to a parallel right-linear production $V_i \to (x \| y) V_j$,

where $V_i, V_j \in V$ for some i, j and $x, y \in T$.

Example 11.11

Let $G_{sp} = (V, T, P, S)$ be a parallelisable string-based SP-grammar with $T = \{a \| a\}$, $V = \{S, A, B\}$, S being a start variable, and $P = \{S \to (a \| a)S, S \to (a \| a)\}$.

The sequences

$S \Rightarrow (a \| a)$

$S \Rightarrow (a \| a)S \Rightarrow (a \| a)(a \| a)$

$S \Rightarrow (a \| a)S \Rightarrow (a \| a)(a \| a)S \Rightarrow (a \| a)(a \| a)(a \| a)$ are some derivations of G_{sp}.

$L(G_{sp}) = \{(a \| a)^m : m > 0\}$ is the language generated by G_{sp}, and the corresponding parallel regular expression is $r = (a \| a)^+$.

Example 11.12

Consider a right-linear grammar $G_{sp} = (V, T, P, S)$ with $T = \{a, b, c\}$, $V = \{S, A, B, C\}$, $P = \{S \to aA, A \to bB, B \to c\|S, B \to c\}$. We have an equivalent parallel right-linear grammar G_{sp} with $T = \{a, b, c, c\|a\}$, $V = \{S, A, B\}$, $P = \{S \to aA, A \to bB, B \to (c\|a)A, B \to c\}$, and S being a start variable.

The sequences

$$S \Rightarrow aA \Rightarrow abB \Rightarrow abc$$

$$S \Rightarrow aA \Rightarrow abB \Rightarrow abc\|a.A \Rightarrow abc\|abB \Rightarrow abc\|abc$$

are some derivations of G_{sp}.

$L(G_{sp}) = \{(abc)^{n\oplus} : n > 0\}$ is the language generated by G_{sp}, and the corresponding parallel regular expression is $r = (abc)^{\oplus}$.

Theorem 11.2

Let $G_{sp} = (V, T, P, S)$ be a parallelisable string-based SP-parallel right-linear grammar. Then $L(G_{sp})$ is a parallelisable string-based SP-regular language.

Proof. Consider G_{sp} a parallelisable string-based SP-parallel right-linear grammar where T, V, and P are the collection of terminals, variables, and production rules, respectively. S is a start variable. Let $V = \{q_0, q_1, q_2, ..., q_n\}$, $T = \{a_1, a_2, a_3, ..., a_n\}$, $S \in V$, and

$$P = \{q_i \to a_j q_k, q_i \to (a_j \| a_k) q_j\}$$

We construct a branching automaton $\mathbf{B} = (Q, \Sigma, \delta_{seq}, \delta_{fork}, \delta_{join}, S, E)$ for a parallelisable string-based SP-parallel right-linear grammar G_{sp} as given below.

We have two possible words of the form

$$x = x_1 x_2 x_3 ... x_n, x_i \in \Sigma^{\oplus}$$

$$y = y_1 \| y_2 \| y_3 \| ... \| y_n, y_j \in \Sigma^*$$

for some i, j where each $x_i = a_1 \| a_2 \| a_3 \| ... \| a_n$ and $y_j = a_1 a_2 a_3 ... a_n$ each for $n \geq 0$. Transitions are defined as:

$$P = \delta_{seq}(q_i, a_j) = q_k \text{ if } q_i \to a_j q_k$$

$$\delta_{join} \delta_{seq} \delta_{fork}(q_i, a_j \| a_k) = q_j \text{ if } q_i \to (a_j \| a_k) q_j$$

The automaton thus constructed will reproduce the derivations by considering each of the productions in terms of transitions.

The initial state of the automaton will be labelled q_0, and for each variable V_i, there will be a state labelled q_i. For each production,

$$q_0 \to a_1 q_i$$

$$q_i \to a_2 q_i$$

$$q_n \to q_n$$

or

$$q_0 \to a_1 \| q_i$$

$$q_i \to a_2 \| q_i$$

$$q_n \to a_n$$

can be combined as

$$q_0 \Rightarrow^* a_1 a_2 a_3 \dots a_{n-1} q_n$$

or

$$q_0 \to (a_1 \| a_2 \| a_3 \| \dots \| a_i) q_n$$

The branching automaton will have transitions to connect q_0 and q_n, where δ is defined as

$$\delta^*(q_0, a_1 a_2 a_3 \dots a_i) = q_i$$

$$\delta_{join}\delta_{seq}\delta_{fork}(q_0, a_1 \| a_2 \| a_3 \| \dots \| a_i) = q_i. \text{ That is,}$$

$$\delta_{join}\delta_{seq}\delta_{join}(q_0, a_1 \| a_2 \| a_3 \| \dots \| a_i) = \delta_{join}\delta_{seq}(\{q_{01}, q_{02}, q_{03}, \dots, q_{0i}\}, a_1 \| a_2 \| a_3 \| \dots \| a_i)$$

$$= \delta_{join}\delta_{seq}(\{(q_{01}, a_1), (q_{02}, a_2), \dots, (q_{0i}, a_i)\})$$

$$= \delta_{join}(\{q_{i1}, q_{i2}, q_{i3}, \dots, q_{ii}\})$$

$$= q_i$$

where $\{q_{01}, q_{02}, q_{03}, \dots, q_{0i}\}, \{q_{i1}, q_{i2}, q_{i3}, \dots, q_{ii}\} \in M_n(Q)$.

By the above construction, if we consider a string $z \in L \subseteq SP\ (\Sigma)$, then either $z \in \Sigma_p^*$ or $z \in \Sigma_s^\oplus$. Their corresponding transition functions are as follows:

$$\delta^*(q_0, x) = q_n$$

$$\delta_{\text{join}}\delta_{\text{seq}}\delta_{\text{join}}(q_0, y) = q_n$$

If q_n is a final state, then z is accepted.

Conversely, assume that z is accepted by branching automaton **B**. Because of the way in which **B** is constructed to accept z, the automaton has to pass through a sequence of states $q_0, q_1, q_2,...,q_n$, using paths labelled $a_1, a_2, a_3,..., a_n$. With, $x = x_1x_2x_3...x_n$, $x_i \in \Sigma^{\oplus}$, $y = y_1\|y_2\|y_3\|...\|y_n$, $y_j \in \Sigma^*$, and the derivation rules as given above. Hence, x, $y \in L(G_{\text{sp}})$ and the theorem is proved.

Example 11.13

Let $G_{\text{sp}} = (V, T, P, S)$ be a parallelisable string-based SP-grammar with $T = \{a\|b\|a\}$, $V = \{S, A, B\}$, $P = \{S \to (a\|b\|a)S, S \to (a\|b\|a)A, A \to (a\|b\|a)\}$, and S being a start variable. The sequences of some derivations are given below.

$S \Rightarrow (a\|b\|a)A \Rightarrow (a\|b\|a)(a\|b\|a)$

$S \Rightarrow (a\|b\|a)S \Rightarrow (a\|b\|a)(a\|b\|a)A \Rightarrow (a\|b\|a)(a\|b\|a)(a\|b\|a)$

$L(G_{\text{sp}}) = \{(a\|b\|a)^m : m > 1\}$ is the language generated by G_{sp}. Branching automaton for $L(G_{\text{sp}})$ is given in Figure 11.4.

Example 11.14

Consider a parallel right-linear grammar $G_{\text{sp}} = (V, T, P, S)$, where $T = \{a, b, c, c\|a\}$, $V = \{S, A, B\}$, $P = \{S \to aA, A \to bB, B \to (c\|a)A, B \to c\}$, and S is a start variable. Then $L(G_{\text{sp}}) = \{(abc)^{m\oplus} : m > 1\}$ is the language generated by G_{sp}. Branching automaton for $L(G_{\text{sp}})$ is given in Figure 11.5.

Remark. We have introduced parallelism and parallel closure operators only for strings of bounded depth. In other words, every parallel operator increases the width of a string. Even though it is of finite depth, introducing the structure of automaton using *fork* and *join* transitions is complicated.

Hence, branching automaton for parallelisable string-based SP-grammar exists only for parallel-series languages.

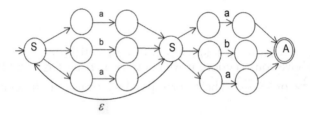

Figure 11.4 Branching automaton for $L(G_{\text{sp}}) = \{(a\|b\|a)^m : m > 1\}$.

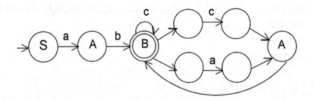

Figure 11.5 Branching automaton for $L(G_{ap}) = \{(a\|b\|c)^{m\oplus} \cdot m > 0\}$

Theorem 11.3

If L is a parallelisable string-based SP-regular language, there exists a parallel right-linear SP-grammar G_{sp} such that $L = L(G_{sp})$.

Proof. Let $\mathbf{B} = (Q, \Sigma, \delta_{seq}, \delta_{fork}, \delta_{join}, S, E)$ be a branching automaton that accepts L. We assume that $\Sigma = \{a_1, a_2,..., a_n\}$ and $Q = \{q_0, q_1, q_2, q_3,..., q_n\}$. Construct a parallel right-linear SP-grammar $G_{sp} = (V, T, P, S)$ with $V = q_0$, $q_1, q_2, q_3,..., q_n$ and $S = q_0$.

For each transition,

$$\delta_{seq}(q_i, a_j) = q_k$$

$$\delta_{join}\delta_{seq}\delta_{join}(q_i, a_j\|a_k) = q_k$$

on **B**, we have productions P of the form

$$q_i \rightarrow a_j q_k$$

$$q_i \rightarrow (a_j\|a_k)q_k$$

We first show that G_{sp} defined in this way can generate every string in L.
Consider $z \in L \subseteq SP(\Sigma)$, then

$$x = x_1 x_2 ... x_n \text{ for } x_i \in \Sigma^\oplus$$

$$y = y_1\|y_2\|...\|y_n \text{ for } y_j \in \Sigma^*$$

Here, each $x_i = a_1\|a_2\|a_3\|...\|a_n$ and $y_j = a_1 a_2 a_3 ... a_n$ for $n \geq 0$.
For B to accept the strings x_i and y_j, it must make moves through the following transitions $\delta(q_0, a_1) = q_1, \delta(q_1, a_2) = q_2,..., \delta(q_{n-1}, a_n) = q_n$ or

$$\delta_{join}\delta_{seq}\delta_{fork}(q_0, a_1\|a_2\|a_3\|...\|a_n) = \delta_{join}\delta_{seq}(\{q_{01}, q_{02}, q_{03},..., q_{0n}\},$$
$$a_1\|a_2\|a_3\|...\|a_n)$$

$$= \delta_{join}\delta_{seq}(\{(q_{01}, a_1), (q_{02}, a_2),..., (q_{0n}, a_n)\})$$

$$= \delta_{join}(\{q_{n1}, q_{n2}, q_{n3},..., q_{nn}\})$$

$$= q_n$$

where $\{q_{01}, q_{02}, q_{03},\ldots, q_{0n}\}, \{q_{n1}, q_{n2}, q_{n3},\ldots, q_{nn}\}\in M_n(Q)$. From the above transitions, we get

$$\delta^*(q_0, x) = q_n$$

$$\delta_{\text{join}}\delta_{\text{seq}}\delta_{\text{fork}}(q0, y) = q_n$$

By construction, the grammar will have a production for each of these δ's. Therefore, we can obtain the derivations for x_i and y_j as

$$q_0 \Rightarrow a_1 q_1$$

$$\Rightarrow a_1 a_2 q_2$$

$$\Rightarrow a_1 a_2 a_3 q_3$$

$$\Rightarrow^* a_1 a_2 a_3 \ldots a_{n-1} q_{n-1}$$

$$\Rightarrow a_1 a_2 a_3 \ldots a_{n-1} a_n$$
and
$$q_0 \Rightarrow a_1 \| q_1$$

$$\Rightarrow a_1 \| a_2 \| q_2$$

$$\Rightarrow a_1 \| a_2 \| a_3 \| q_3$$

$$\Rightarrow a_1 \| a_2 \| a_3 \| \ldots \| a_{n-1} \| q_n$$

$$\Rightarrow a_1 \| a_2 \| a_3 \| \ldots \| a_n$$

To generate x and y, we can have the productions:

$$q_0 \rightarrow x_i \| q_i$$

$$q_i \rightarrow x_i$$

or

$$q_0 \rightarrow y_i q_i$$

$$q_j \rightarrow y_j$$

with the grammar G_{sp} and each x_i and y_j is generated by the above construction. Hence, $z \in L(G_{sp})$.

Conversely, if $z \in L(G_{sp})$, then its derivation must have the form given above. This implies that

$$\delta^*(q_0, x) = q_n$$

or

$$\delta_{join}\delta_{seq}\delta_{fork}(q_0, y) = q_n$$

completing the proof.

Theorem 11.4

If L is a parallelisable string-based SP-regular language, there exists a parallel left-linear SP-grammar G_{sp} such that $L = L(G_{sp})$.

Proof. Consider $L \subseteq SP\ (\Sigma)$ a parallelisable string-based SP-regular language. By Theorem 9.6.10, there exists a parallel right-linear grammar G_{sp}. Given any parallel left-linear SP-grammar with production rules

$$V_i \to V_j x \text{ or } V_i \to x$$

where $V_i, V_j \in V$ and $x \in T$.

We construct parallel right-linear SP-grammar G'_{sp} by replacing every production of G_{sp} with $V_i \to x^R V_j$ or $V_i \to x^R$,

respectively, where x^R denotes the reverse of x, G'_{sp} is right-linear, and $L(G'_{sp})$ is regular by Theorem 11.2. Also, the reverse of any regular language is regular.

Therefore, $L(G'_{sp})^R = L(G)$ and hence proved.

Theorem 11.5

A parallelisable string-based SP-language L is regular iff there exists a SP-regular grammar G_{sp} such that $L = L(G_{sp})$.

Proof. Combining Theorems 11.4 and 11.3, we arrive at the equivalence of regular languages and regular grammars.

Theorem 11.6

For $L \subseteq SP\ (\Sigma)$, the conditions given below are equivalent:

1. L is parallelisable string-based SP-regular.

2. $L = L(r)$ for a parallel regular expression r.
3. L is generated by a parallel SP-regular grammar G_{sp}.

Proof. (2) $\Rightarrow$ (1) follows from Theorem 11.1.
(1)$\Rightarrow$(3) follows from Theorem 11.5.

11.7 CONCLUSIONS

This chapter discussed the parallelisable string-based SP-languages as the union of series-parallel languages and parallel-series languages. It defined the parallel regular expression to express the languages, and also the grammar to generate parallelisable string-based SP-languages.

Recognition of branching automaton through parallelisable string-based SP-parallel regular grammar and parallel regular expression has also been discussed in this chapter.

11.8 APPLICATIONS

- For exploiting the parallelism inherent in divide and conquer algorithms on shared memory multiprocessors, fork/join parallelism is used.
- If components of electrical circuits are connected in series in some portions and parallel in others, it is not possible to proceed a *single* set of rules to every portion of that circuit. Instead, it will be good to find which parts of that circuit are series and which parts are parallel, and then apply series and parallel rules selectively to complete the process in short.

11.9 FUTURE SCOPE

Equivalence properties, regularity properties, emptiness, decidable conditions, etc., are needed to be explored for the parallelisable string-based SP-languages. All the properties are to be applied in a suitable scenario to find their validity.

REFERENCES

1. J. Hopcroft and J. Ullman, *Introduction to Automata Theory, Languages and Computation*, Addison-Wesley, Reading, MA, 1979.
2. W. Thomas, Languages, automata and logic, In: G. Rozenberg, A. Salomaa (Eds.), *Handbook of Formal Languages*, Springer, Berlin, 1997, 389–455.

3. D. Beauquier and J.E. Pin, Languages and scanners, *Theoretical Computer Science*, 84 (1991) 3–21.
4. S. Eilenberg, *Automata, Languages and Machines*, Vol. A (1974) Academic Press, New York.
5. K. Lodaya and P. Weil, Series parallel posets: algebra, automata and languages, In: M. Morvan, C. Meinel and D. Krob (Eds.), *Proceedings of STACS (Paris 98), LNCS 1373*, Springer, Berlin, 1998, 555–565.
6. K. Lodaya and P. Weil, Series-Parallel languages and the bounded-width property, *Theoretical Computer Science*, 237 (2000) 347–380.
7. K. Lodaya and P. Weil, A Kleene iteration for parallelism, *Foundations of Software Technology and Theoretical Computer Science*, 1530 (1998) 355–366.
8. K. Lodaya and P. Weil, Rationality in algebras with a series operation, *Information and Computation*, 171 (2001) 269–293.
9. I.M. Havel, Finite branching automata, *Kybernetika* 10 (1974) 281–302.
10. N. Bedon, Logic and branching automata, *Logical Methods in Computer Science* 11 (2015) 1–38.
11. D. Kuske, Infinite series-parallel posets: logic and languages, *ICALP 2000, Lecture Notes in Computer Science* Vol. 1853, Springer-Verlag, 2000 648–662.
12. A. Amazigh and N. Bedon, Equational theories of scattered and countable series-parallel posets, *Developements in Language Theory*, 12086 (2020) 1–13.
13. B. Kemper and M. Mandjes, Mean sojourn times in two-queue fork-join systems: bounds and approximations, *OR Spectrum* 34, (2012) 723–742.

Chapter 12

Detection of disease using machine learning

Bansi Patel, DhruvinSinh Rathod, and Samir Patel

Pandit Deendayal Energy University

CONTENTS

12.1 INTRODUCTION

As we know, there are too much queues in hospitals day by day due to different diseases and doctors requires different testing methods for the detection of disease that the patient has. And as we know, testing and reports require around 2–3 hours per patient and again they have to see the doctor, and it requires additional time of doctor as well as patient. So, as we know, there are some nations having large population in the world and some of them are having very less doctors, resulting in less testing rate in such nations. So our overall idea about this work is to provide web-based solution that helps the doctors to increase testing for various diseases and helps the world to be disease-free as far as possible. Sometimes, it is difficult for even doctors to detect disease more accurately by just observing the patient X-ray or microscopic image.

Achieving this thing in the current scenario is possible due to evolution in the field of AI and computational capability and availability of data. As we know, nowadays, digital data are easy to capture and are available. It has become possible due to easy Internet availability and increased computation power of computers along with communication technologies ever expanding such as 4G and 5G. Cloud computing technologies are available along with AutoML tools, which provide automatic finding of best hyper-parameter and best model. So we can develop AI-enabled solutions, which can help the doctors to detect disease from the available information (e.g., X-rays, CT scan images, microscopic data, and pathological data). We have developed various AI- and ML-based algorithms which are able to detect various diseases such a corona [1] and malaria [2]. Here, all the algorithms are trained with required input data such as chest X-ray images of corona patient. For malaria detection, we use the microscopic image of the patient blood, to predict the outcome of each of the diseases. We are currently able to achieve 83% accuracy in corona detection and 83% accuracy in malaria detection.

12.2 TECHNIQUES APPLIED

12.2.1 Overall system architecture

Our main idea is to develop an AI-based smart system that helps doctors and patients in many ways.

In this work, we develop AI-based algorithms, which will be used to detect various diseases by just giving the required input parameter such as X-ray or other parameters such as blood pressure and microscopic image. From this system architecture, we can train the machine for detecting various types of diseases.

After developing our different algorithm, we will integrate it with Firebase online database. As a front-end side, we are going to develop Python- and Flask-based website [3] through which different doctors can provide inputs for disease testing for different patients.

12.2.2 Working of overall system

As shown in Figure 12.1, doctors will use the website to provide input as well as to get the output. For example, doctors want to detect whether the

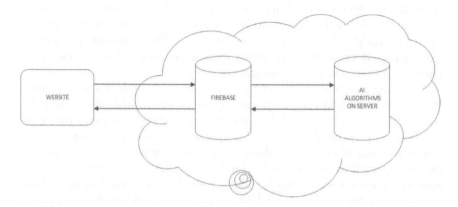

Figure 12.1 General structure.

patient has corona or not; they will feed the X-ray image of the patient through the website.

After uploading the image, it will be sent to the Firebase online database. Here, Firebase will use asynchronous method to send data or image to AI algorithm that resides at server side.

Once the server got the image, it will have pre-trained AI algorithms for different diseases. It will load the model in memory and will provide data or image as input to the model

Once the model gets executed, the server will collect the output and again send the data to Firebase. After receiving data, Firebase sends the predicted data to the requested doctor. Here, Firebase database and server act as a cloud. Then to develop the sever-side AI algorithms, we use data mining techniques and algorithms such as deep learning-based CNN [4-6], vanilla neural network [7,8], VGG-19, and ensemble learning.

12.3 GENERAL ARCHITECTURE OF AI/ML

For each and every disease, we have used this AI/ML algorithm architecture to get good accuracy (Figure 12.2). The steps that we have followed to achieve the good accuracy and find good models are as follows:

1. Gathering of dataset and analysis
2. Preprocessing of data
3. Splitting of data into training and test datasets
4. Creating AI/ML model and training on training dataset and hyper-parameter tuning
5. Evaluating the model using test dataset.

12.3.1 Corona detection

i. Gathering dataset and analysis: To gather the dataset for corona, we use the dataset available online with X-ray images of different patients with and without corona.

Here, each image has a different shape. Some images are in RGB format (3 channels), while some images are in black and white format (1 channel) [9].

Here we have the following number of images for each disease. So, our dataset has various classes as outcome Table 12.1.

ii. Data preprocessing: As shown in Figure 12.3, from analysis, the dataset we have is multiclass, but we want to check only for corona and we want to treat this problem as binary class instead of multiclass. If the model predicts negative for corona, then it means the patient will have disease from other remaining classes that we have treated as single class.

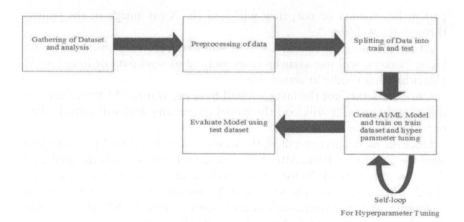

Figure 12.2 General architecture of AI/ML Ref. [10].

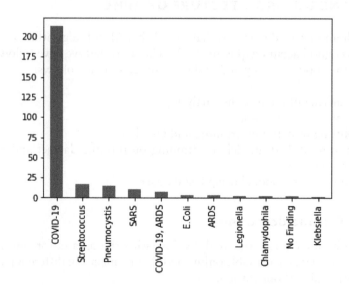

Figure 12.3 Analysis of dataset of COVID-19 [10].

So as preprocessing, we first converted the given data into binary class by considering COVID-19 and COVID-19, ARDS as single class and all other diseases, as second class. After that, we check the distribution of the dataset (Figure 12.4). We have 199 images of corona-positive patients and 58 images of corona-negative patients. So, it indicates that we have a class imbalance problem. To solve this problem, we use the technique called oversampling to make both the class balanced. The result we get after oversampling is shown in Figure 12.5.

Table 12.1 Disease and total outcomes from the dataset [10]

Disease	Outcomes
COVID-19	213
Streptococcus	17
Pneumocystis	15
SARS	11
COVID 19, ARDS	8
E. Coli	4
ARDS	4
Legionella	2
Chlamydophila	2
No finding	2
Klebsiella	1

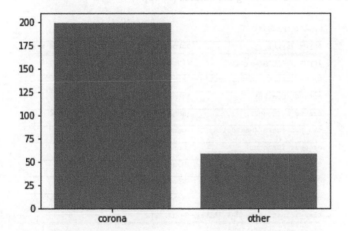

Figure 12.4 Distribution of the dataset [11-13].

To perform oversampling, we generated new images by flipping images up, down, left, and right and now we have 221 images of each class. After that, we convert all the images into 128×128 size and RGB channel (three channel) for both the classes, so that the model can be trained faster with 128×128 image size. At output column, we converted output into one-hot vector encoding form.

i. Train and test data split: We then divide our data into two parts: training and test sets, of which we have 33% for testing and the remaining for training.
ii. Create machine learning model: To create ML model, we use convolutional neural network. We then created models which are identical, but trained with different numbers of epochs.

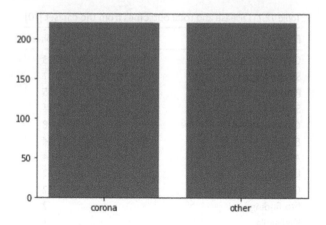

Figure 12.5 After oversampling the dataset [11-13].

```
Model: "model"

Layer (type)                    Output Shape              Param #
=================================================================
input_1 (InputLayer)            [(None, 128, 128, 3)]     0

normalization (Normalization    (None, 128, 128, 3)       7

conv2d (Conv2D)                 (None, 126, 126, 32)      896

conv2d_1 (Conv2D)               (None, 124, 124, 64)      18496

max_pooling2d (MaxPooling2D)    (None, 62, 62, 64)        0

dropout (Dropout)               (None, 62, 62, 64)        0

flatten (Flatten)               (None, 246016)            0

dropout_1 (Dropout)             (None, 246016)            0

dense (Dense)                   (None, 2)                 492034

classification_head_1 (Activ    (None, 2)                 0
=================================================================
Total params: 511,433
Trainable params: 511,426
Non-trainable params: 7
```

Figure 12.6 Training model for the corona detection.

As shown in Figure 12.6, two models are trained using our data-set. Here, both models are similar in structure, but we have different hyper-parameters for both of them, so both have different accuracies.

Here, Model 1 has 82% accuracy, while Model 2 has 79% accuracy. To increase the accuracy, we apply the concept of ensemble learning. For applying ensemble learning, we merge this two-model output on which we have done the voting and it increases our accuracy to 83%. In order to find best parameters we use Autokeras tool [14,15] to derive the optimal parameter for model. Here, we use 128×128 size images as input for the model and we have around 442 images from which we use 33% as testing dataset. The number of epochs used to achieve good accuracy is 20–30. The optimal parameters found for this model by the AutoKeras are as follows:

- kernel size=3
- num blocks=1
- num layers=2
- dropout1 value=0.25
- dropout2 value=0,50
- filter1=32
- filter2=64
- optimizer=adam
- learning rate=0.001.

iii. Evaluate model: To evaluate the model, we use accuracy metrics as shown in equation 12.1.

$$Accuracy = (\text{Total no. of correct predictions})/(\text{Total no. of instances}) \quad (12.1)$$

We use these accuracy metrics on the test dataset and obtained 83% accuracy for testing X-ray image data.

12.3.2 Malaria detection

i. Gathering dataset and analysis: To gather the dataset, we use online malaria detection dataset. The dataset we use has two parts: parasite and uninfected.

The parasite part of the dataset contains the images of patient with malaria disease and uninfected part of the dataset contains non malaria instances (see Figure 12.7). The histogram shows that the dataset that we use does not have a class imbalance problem.

ii. Preprocessing: The data we use don't have any imbalance and don't require anything to perform.

iii. Train and split data: For splitting the dataset, we have divided our dataset into 50% for training and 50% for testing.

iv. Create ML model: The images were first converted into 224×224 pixels for processing with the channel size of 3. There are total number of 5 epochs performed to achieve this accuracy. We use 1,000 images to train this model, and for testing, we use 250 images. In Figure 12.8, you can see that training vs. value loss in the model.

v. Evaluate model: To evaluate the model, we use accuracy metrics as shown in equation (12.1). Here we use the VGG-19 algorithm to train the model for malaria detection. This model gives us the accuracy of 83%.

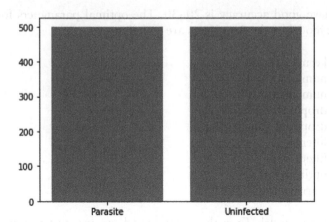

Figure 12.7 Malaria balanced dataset.

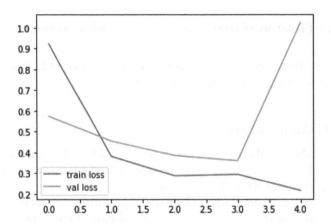

Figure 12.8 Train vs. value loss in the model.

Layer (type)	Output Shape	Param #
input_1 (InputLayer)	(None, 224, 224, 3)	0
block1_conv1 (Conv2D)	(None, 224, 224, 64)	1792
block1_conv2 (Conv2D)	(None, 224, 224, 64	36928
block1_pool (MaxPooling2D)	(None, 112, 112, 64)	0
block2_conv1 (Conv2D)	(None, 112, 112, 128)	73856
block2_conv2 (Conv2D)	(None, 112, 112, 128)	147584
block2_pool (MaxPooling2D)	(None, 56, 56, 128)	0
block3_conv1 (Conv2D)	(None, 56, 56, 256)	295168
block3_conv2 (Conv2D)	(None, 56, 56, 256)	590080
block3_conv3 (Conv2D)	(None, 56, 56, 256)	590080
block3_conv4 (Conv2D)	(None, 56, 56, 256)	590080
block3_pool (MaxPooling2D)	(None, 28, 28, 256)	0
block4_conv1 (Conv2D)	(None, 28, 28, 512)	1180160
block4_conv2 (Conv2D)	(None, 28, 28, 512)	2359808
block4_conv3 (Conv2D)	(None, 28, 28, 512)	2359808
block4_conv4 (Conv2D)	(None, 28, 28, 512)	2359808
block4_pool (MaxPooling2D)	(None, 14, 14, 512)	0
block5_conv1 (Conv2D)	(None, 14, 14, 512)	2359808
block5_conv2 (Conv2D)	(None, 14, 14, 512)	2359808
block5_conv3 (Conv2D)	(None, 14, 14, 512)	2359808
block5_conv4 (Conv2D)	(None, 14, 14, 512)	2359808
block5_pool (MaxPooling2D)	(None, 7, 7, 512)	0
flatten_1 (Flatten)	(None, 25088)	0
dense_1 (Dense)	(None, 2)	50178

Total params: 20,074,562
Trainable params: 50,178
Non-trainable params: 20,024,384

Figure 12.9 Model for malaria detection [7].

12.4 EXPERIMENTAL OUTCOMES

The outcome of the project is measured in the form of accuracy of the model and to calculate the accuracy we are using accuracy metrics. We have currently finished the detection of two diseases, for which the accuracies are listed below:

- Coronavirus detection using X-ray: 83% accuracy.
- Malaria detection: 83% accuracy.

12.5 CONCLUSIONS

In the current study, we have proposed an AI-enabled approach for the detection of diseases such as corona and malaria. Corona and malaria disease detection models are based on CNN. The accuracy metrics are used here to evaluate all the models' performance on testing dataset. All the hyper-parameters are found based on different experiments as well as using different tools and libraries such as h2o and AutoKeras.

REFERENCES

1. "Malaria cell image classification using deep learning," *Int J Recent Technol Eng*, vol. 8, no. 6, pp. 5553–5559, Mar. 2020. Doi: 10.35940/ijrte. f9540.038620.
2. "Researchers use AI to detect COVID-19," Imaging Technology News, 2020. [Online]. Available from: https://www.itnonline.com/content/researchers-use-ai- detect-covid-19. (Accessed: 01 August 2020).
3. J. Waring, C. Lindvall, and R. Umeton, "Automated machine learning: Review of the state-of-the-art and opportunities for healthcare," *Artif Intellig Med*, vol. 104, p. 101822, Apr. 2020. Doi: 10.1016/j.artmed.2020.101822.
4. M. Xin and Y. Wang, "Research on image classification model based on deep convolution neural network," *J Image Video Proc*, vol. 2019, no. 1, Feb. 2019, Doi: 10.1186/s13640-019-0417-8.
5. H. Jin, Q. Song, and X. Hu. "Auto-Keras: An efficient neural architecture search system," vol. 3, Mar. 2019, https://arxiv.org/abs/1806.10282v3.
6. Md. Z. Alam, M. S. Rahman, and M. S. Rahman, "A random forest based predictor for medical data classification using feature ranking," *Inform Med Unlock*, vol. 15, p. 100180, 2019. Doi: 10.1016/j.imu.2019.100180.
7. M. Mateen, J. Wen, S. Song, and Z. Huang, "Fundus image classification using VGG-19 architecture with PCA and SVD," *Symmetry*, vol. 11, no. 1, p. 1, Dec. 2018, Doi: 10.3390/sym11010001.
8. M. Poostchi, K. Silamut, R. J. Maude, S. Jaeger, and G. Thoma, "Image analysis and machine learning for detecting malaria," *Translat Res*, vol. 194, pp. 36–55, Apr. 2018. Doi: 10.1016/j.trsl.2017.12.004.

9. S.-J. Lee, T. Chen, L. Yu, and C.-H. Lai, "Image classification based on the boost convolutional neural network," *IEEE Access*, vol. 6, pp. 12755–12768, 2018. Doi: 10.1109/access.2018.2796722.

10. W. Rawat and Z. Wang, "Deep convolutional neural networks for image classification: A comprehensive review," *Neural Comput*, vol. 29, no. 9, pp. 2352–2449, Sep. 2017. Doi: 10.1162/neco a 00990.

11. N. Rout, D. Mishra, and M. K. Mallick, "Handling imbalanced data: A survey," In *Advances in Intelligent Systems and Computing*, Springer Singapore, pp. 431–443, 2017. DOI:10.1007/978-981-10-5272-9_39

12. R. Jagali, "A reversible data hiding method with contrast enhancement for medical images by preserving authenticity," *GRD J- global Res Develop J Eng*, vol. 1, no. 9, Aug. 2016 ISSN: 2455-5703.

13. A. Singh and A. Purohit, "A survey on methods for solving data imbalance problem for classification," *Int J Conformity Assessment*, vol. 127, no. 15, pp. 37–41, Oct. 2015. Doi: 10.5120/ijca2015906677.

14. Q. Gu, Z. Cai, L. Zhu, and B. Huang, "Data mining on imbalanced data sets," *Presented at the 2008 International Conference on Advanced Computer Theory and Engineering (ICACTE)*, Dec. 2008. Doi: 10.1109/icacte.2008.26.

15. Corona X ray image dataset. https://github.com/ieee8023/covid-chestxray-dataset.

Chapter 13

Driver drowsiness detection system using eye tracing system

Saketram Rath and Debani Prasad Mishra
International Institute of Information Technology Bhubaneswar

CONTENTS

13.1 INTRODUCTION

Accidents due to vehicles is on incessant rise despite the several measures taken by the local authorities to curb them down. Driver drowsiness and fatigue have been the most pertaining causes of these accidents [1]. Drowsiness is a subconscious state of mind caused due to lack of sleep. When a driver starts feeling drowsy, they are not alert enough to take the responsibility of driving a vehicle during tricky parts of the day (e.g., twilight and night period). Since the start of introduction of cheaper Internet options and the availability of Internet in the remote locations of the world, almost 90% of drivers are equipped with a smartphone that has GPS, location, and cameras on both the front and back portion of the mobile [2]. This chapter presents the idea of real-time drowsiness detection of drivers with eye tracking, which requires a very basic camera which is capable of capturing images of 320×240 pixels per inch density at 110 frames per second. We mainly deal with the physical aspect of fatigue, which is generally noticeable from an individual with certain arrangement of their eyes. This criterion can be determined with the help of Viola-Jones algorithm. The Viola-Jones cascade amplifier is utilized to determine whether the eyelids of the driver is open or closed and trigger the said alarm accordingly [3]. The Viola-Jones method along with PERCLOS method results in 99% accuracy in eye movement and blinking detection [4]. The drowsiness detection is carried out in the following methods such as eye blinking frequency; yawning frequency; eye gaze movements; facial expressions; and head movements.

Eye movements such as blinking, eye fixation, and pupil size during driver fatigue can be utilized in the driver fatigue detection system with the help of the Haar cascade classifier [5]. Since the system is visionary in nature, it is designed to be non-intrusive in nature so that it may not hamper the driver's attention at the period of driving, but can be used simultaneously to monitor the alertness of the driver. The open-CV module of data science is used in the model in order to perform all the live tracking of the features, since it is equipped with all the features of the Viola-Jones algorithm [6].

13.2 LITERATURE REVIEW

The drowsiness monitoring system has been implemented by various luxury car brands as an accessory. In this present age of increasing vehicle population, it has become a necessity. This chapter mentions about a cheaper alternative that can be accessed by any driver with a smartphone with a well-functioning front camera. It presents the use of the Viola-Jones algorithm as an alternative to such complicated methods. The algorithm just requires a proper functioning camera that is able to capture images in grayscale format. The algorithm is then able to identify several physical features of the face by segregating them in the form of rectangle boxes and can monitor their movement continuously and effectively for a longer period of time [7]. This feature is called percentage of eye closure (PERCLOS) analysis feature. In visual feature analysis, PERCLOS (percentage of eye closure) is the attentiveness level. PERCLOS imitates slow eyelid closure which is the proportion of total time of driver's eyelid is closed for 80% or more duration of time. There is a specific range of values for the PERCLOS method where the driver is deemed fit for driving. Any deviation from the set of values will prompt the system to alert the driver by generating an alarm accordingly [8]. This method is a cheaper and more accurate alternative to the ones implemented earlier. This method can be used as an application that can be accessed by all using a smartphone and can be implemented in any type of vehicles starting from two wheelers to heavy vehicles such as trucks, railway locomotives, and airplane cockpits [9]. Several other methods for drowsiness monitoring system were implemented. The EEG method using the ears is the most tedious one, even though it's very accurate [10–12]. There are also many implementations using the deep curvature and sparse learning method with pattern tracing, but due to its complexity, it is very less preferred in comparison with other methods [13]. Several wearable devices equipped with oximeter and capable of tracking sleep cycles of the drivers have also been introduced. These devices monitored the heart rate, pulse rate, and several other bodily functions of the human body, but these were not equipped enough to handle real-time tracking of drowsiness. These devices are often very fragile, and any carelessness can affect its functionality [14,15].

This chapter is further structured to discuss the research methods followed by results and observations. The observations are then succeeded by the conclusions followed by the bibliography.

13.3 RESEARCH METHODS

The Viola-Jones method is the very first algorithm with real-time face capturing and tracking. There are three main procedures that can complete the smooth face tracking feature without any hassle, along with complete accuracy and very less delay. AdaBoost for function collection and an attention-grabbing cascade for optimal allocation of computing capital. This chapter provides a complete description of the algorithm here [16].

13.3.1 Video acquisition using web camera

This is the initial step of the process where the front camera of the setup is turned on. The videos are captured in the feed of the driver of the vehicle in real time. The acquired video is now processed and then divided into several frames. To take live video as its input, this module is used and translated into a set of images/frames, which are then stored. Likewise, in the algorithm, when the code is run, the webcam is initially triggered, takes the live feed, and turns it into appropriate frames, i.e., 10 frames. For this process, various functions from the open-CV modules are implemented [17]. The flowchart represented in Figure 13.1 shows the working of the video acquisition from the web camera.

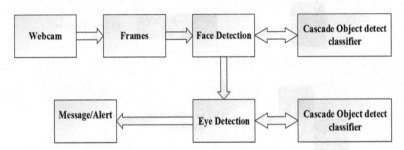

Figure 13.1 Flowchart of real-time video acquisition using open-CV module by accessing the webcam/front camera.

The flowchart in Figure 13.1 denotes the working of video acquisition of the user from the user perspective camera. The videos are recorded in real time and processed in the form of continuous images. The images are then further processed to undergo.

13.3.2 Non-separable case

13.3.2.1 Haar cascade algorithm

The Haar cascade is a machine learning-based algorithm based on extracting important features from any image or audio source. It is based on the concept of features and was proposed by P. Viola and M. Jones in their paper in the year 2001 [18]. The four main parts of this algorithm are Haar feature selection, integral images creation, AdaBoost training, and cascading classifiers.

The very first step of this algorithm is to collect features from the given sample of image or the video. A Haar feature considers adjacent rectangular regions at a specific location in a detection window, sums up the pixel intensities in each region, and calculates the difference between these sums [19]. Edge features: Edges are the basic feature of an image identified with a very sharp pixel difference. This feature detects the facial boundaries of the individual's face, thus differentiating from the rest of the body.

Figure 13.2 explains the working of Haar feature selection. The features are detected in three subsequent methods. Line features: Line features via corresponding direction detectors, the lines in various directions are detected and then fused into one edge image. The lines of the training samples are represented and processed with a chain code in the training

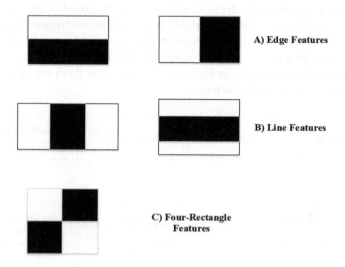

A) Edge Features

B) Line Features

C) Four-Rectangle Features

Figure 13.2 Methods of facial feature extraction using Haar feature selection.

process. Four-rectangle method: This is the most important part of feature extraction that is essential for extracting features such as lips, eyes and nose whose movements are essential in determining the drowsiness state of driver. The edge features detected are then divided into four more sections in the form of rectangles that individually detect the important facial features, thus completing the feature extraction. Since most of these features remain irrelevant, we implement the AdaBoost algorithm. The AdaBoost algorithm is a very powerful boosting algorithm in the field of data science. It focuses on classification of problems and converts a set of weak classifiers into very strong ones. In our approach, we use AdaBoost to select very best features and train them accordingly. This algorithm constructs a "strong" classifier as a linear combination of weighted simple "weak" classifiers [20]. You can see this in the video below, in action. This distinction is then contrasted with a taught threshold that distinguishes objects from non-objects. Since each Haar feature is just a "weak classifier" (its level of identification is marginally higher than random guessing), a large number of Haar features are required to define an object with appropriate precision and are thus arranged to form a strong classifier into cascade classifiers [21].

A strong classifier generated by the AdaBoost algorithm consists of each point. On a sequential (stage-by-stage) basis, an input is evaluated. If a classifier for a particular stage yields a negative result, the input is automatically discarded. The input is redirected to the output if the output is positive. The input is transmitted to the next step. This multistage approach, according to Viola and Jones (2001), allows for the creation of simplified classifiers that can then be used to easily reject the most negative (non-face) input while taking more time on positive (face) input [22].

From Figure 13.3, the working of cascade classifier is explained. The classifier algorithm takes multiple images of the user as input from the web camera. The classifier with the help of its algorithm then detects the presence of all the essential facial features. If all the facial features are present, then the image is further processed, or else it is discarded from further usage.

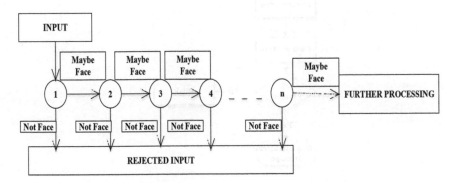

Figure 13.3 Continuous face detection using cascade classifier.

13.3.3 Flowchart representation of the procedure

The flowchart in Figure 13.4 represents the methodology implemented in this chapter and the algorithm used for the purpose. The first step involves the detection of the face from the live feed of the front camera. The Haar feature extraction algorithm extracts the facial features from the camera. The cascade classifier then selects the best possible input for eye detection. The eyes are now tracked with the help of mean shift algorithm. The movement of the eyelids are now tracked from the live feed with the help of PERCLOS algorithm. If the eyes remain closed for a longer period of time, then an alarm beep is played as an output. The alarm is loud enough to alert the driver. The trigger input for the alarm can also be implemented in several other methods as well depending on the connectivity and the features that exist on the particular vehicle.

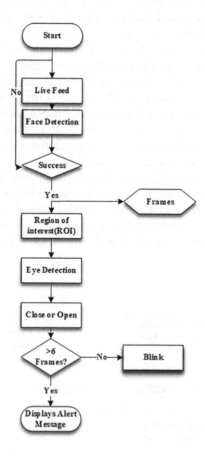

Figure 13.4 Flowchart representation of the methodology of algorithm.

13.4 OBSERVATIONS AND RESULTS

The code for the project uses mainly open-CV module [23]. This module is a derivative of NumPy module and consists of all the functions and operations related to the Haar cascading algorithm and the Viola-Jones algorithm. The model is first trained with a couple of video samples in order to improve its accuracy and efficiency. The model has the access to the camera of your system or the smartphone. It runs the camera, and with the help of Haar feature selection and AdaBoost, all important features of the face are detected. In the following step, the eyes are detected with the help of cascade classifier. With the completion of detection of all the important features, the PERCLOS algorithm is implemented in order to track the blinking of the eyes in real time. If the eye remains closed for more than 1 second, then an alert sound is produced as the output [24].

13.4.1 Face detection

The face detection function obtains one frame at one time among the t frames received from the frame grabber and attempts to detect the face of the car driver in each and every frame. This is accomplished by making use of a pre-defined collection of samples by Haar cascade. The beauty of this algorithm is the use of rectangular properties against multiple pixels. In a rectangular box, total pixels are counted at the start. Then there's the box combinations which are sums of characteristics to identify the face [25].

From Figure 13.5, it can be seen that with the help of cascade classifier, all the important features are being extracted from the live feed of the camera. The facial borders are detected. The borders are now marked under a blue square frame ensuring that the face has been detected and the algorithm proceeds further.

Figure 13.5 Detection of face in real time from the webcam.

13.4.2 Eye Tracking

When the face detection function senses the face of the vehicle driver, the eye detection function aims to identify the eyes of the vehicle driver. This task is accomplished by the Viola-Jones algorithm. Then the pupil, which is RGB in nature, is converted to black and white or binary form or the YCbCr format [26].

In Figure 13.6, the algorithm tracks the eyes from the live feed of the web camera. The eyes are detected by the algorithm and are marked with the help of green border. This also tracks the position of eyelids simultaneously.

13.4.3 Morphological operation

After converting RGB to YCbCr, binarization is done; i.e., it is converted into a binary image that is stored as a single bit of 0 or 1. But tiny background color holes are presented in the binary image. Therefore, the holes must be filled to satisfy the area of interest (ROI) identification. The morphological closing procedure is used to satisfy the tiny context holes [27].

The eyes that are detected by the algorithm are now converted to black and white form from RGB as seen in Figure 13.7. Then the opening of pupils is determined, and the full opening is represented in the form of white spaces in black background as implemented in the PERCLOS algorithm. The intensity of the opening of pupils can be determined from the area occupied by the white spaces. The lesser the area occupied, the more closed are the eyes.

Figure 13.6 Eye tracking using the Viola-Jones algorithm.

Figure 13.7 Conversion of the pupil from RGB form to YCbCr. The pupil openings are depicted in a black background.

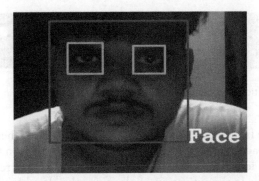

Figure 13.8 Implementation of mean shift algorithm after complete detection of face.

13.4.4 Feature extraction and tracking of pupil

The extraction of the region of iris is required in order to track the movement of the eyes. With the skin tone chosen as the tracking function, we used the mean tracking change algorithm. The mean shift algorithm has an advantage due to its limited computing expense for real-time monitoring. Mean change tracking is an object tracking technique dependent on presence. It employs the mean study of changes to identify the target candidate region, which in terms of intensity distribution has the most comparable appearance to the target model [28].

13.4.5 Mean shift algorithm

Mean shift is a method to find the limit of a density function given the sampled discrete data from that function. In a way, it uses a gradient approximation of non-parametric density. The modes of this density are useful for detecting it. It is an application-independent tool suitable for real-time data analysis. In this chapter, it plays a crucial role in real-time tracking of all the features and then returns the output simultaneously [29].

In Figure 13.8, the working of the mean shift algorithm is denoted; the algorithm detects the consistency of the image and presence of all the required features. The eyes are marked under green square boxes, denoting their presence. The blue box detects the face by detecting the facial boundaries.

13.4.6 Detection of eye closure

After the completion of face tracking, the movement of the eyelids is tracked with the help of percentage of closure of eyes (PERCLOS) algorithm. If the PERCLOS remains more than 90 percent, i.e., the blinking time of eyes exceeds more than one second, then the driver is alerted with the help of an alert sound via the speakers of the smartphone owned by him [30].

Figure 13.9 Detection of eye closure using PERCLOS algorithm.

It can be seen from Figure 13.9 that if the eyes are closed then are occupied by the white spaces are very negligible as implemented from the PERCLOS algorithm. The lesser the area covered, the more the eyes are closed. Thus, this feature helps in the activation of alarm for the algorithm. The accuracy of this feature is up to 93.9%.

13.4.7 Activation of alert sounds

The algorithm detects the positions of eyelids. If the PERCLOS percentage is found to be greater than or equal to 90%, then an impulse output is generated. This impulse output is then utilized by the algorithm as a parameter for the alarm activation.

Alarm is one of the important features of this algorithm. Its intensity is loud, and frequency is high enough to gain the attention of the driver. It is very important to choose to an appropriate sound signal as an alarm. It should not be feeble that it fails to gain the attention of driver, thus resulting in fatalities. In this algorithm, a warning sound is used to alert the driver. This sound has the appropriate intensity in order to gain the attention of drivers.

13.5 CONCLUSIONS

The Viola-Jones method is the building block of this model. The Viola-Jones algorithm is used to classify several faces and places of concern, such as eyes, nose, and lips. Accurate function recognition when a camera with high specs is used, it improves easily. We enforce this device in a vehicle, only the face of the driver, i.e., one face, is identified, and he will get a warning message when the face is identified and recognized as drowsy. Hence, we have successfully designed the required model by implementing the Viola-Jones algorithm to detect the drowsiness of the driver efficiently and effectively.

In this new scientific era, facial tracking is one of the upcoming technologies for vehicular safety. Several features such as dynamic braking and cruise control are being introduced; when combined with the live facial tracking, it can form into a very robust safety feature. The trigger output generated can be used to control various aspects of movement of cars such as lane assists and control on the intensity of braking depending on the drowsiness of the driver. The facial detection is one of the most efficient and affordable safety technologies. With proper implementation, it can be an efficient tool in accident prevention and reduction of injuries and death by a significant number in the near and upcoming future.

REFERENCES

1. E. Johnson, J. M. Abraham, S. Sulaiman, L. Padma Suresh and S. Deepa Rajan, "Study on road accidents using data mining technology," *2018 Conference on Emerging Devices and Smart Systems (ICEDSS)*, Tiruchengode, 2018, pp. 250–252, doi: 10.1109/ICEDSS.2018.8544370.
2. B. Paul, and A. Murti, *Socio-economy of Mobile Phone Ownership in India*; 2016. Doi: 10.1007/978-981-10-1684-4_9.
3. W. Lu and M. Yang, "Face detection based on viola-jones algorithm applying composite features," *2019 International Conference on Robots & Intelligent System (ICRIS)*, Haikou, China, 2019, pp. 82–85, Doi: 10.1109/ICRIS.2019.00029.
4. H. Hu, Y. Zhu, Y. Zhang, Q. Zhou, Y. Feng and G. Tan, "Comprehensive driver state recognition based on deep learning and PERCLOS criterion," *2019 IEEE 19th International Conference on Communication Technology (ICCT)*, Xi'an, China, 2019, pp. 1678–1682, Doi: 10.1109/ICCT46805.2019.8947282.
5. X. Lian, "A face recognition approach based on computer vision," *2020 5th International Conference on Smart Grid and Electrical Automation (ICSGEA)*, Zhangjiajie, China, 2020, pp. 171–175, Doi: 10.1109/ICSGEA51094.2020.00044.
6. S. Anwar, M. Milanova, A. Bigazzi, L. Bocchi and A. Guazzini, "Real time intention recognition," *IECON 2016–42nd Annual Conference of the IEEE Industrial Electronics Society, Florence*, 2016, pp. 1021–1024, Doi: 10.1109/IECON.2016.7794016.
7. A. Acıoğlu and E. Erçelebi, "Real time eye detection algorithm for PERCLOS calculation," *2016 24th Signal Processing and Communication Application Conference (SIU)*, Zonguldak, 2016, pp. 1641–1644, Doi: 10.1109/SIU.2016.7496071.
8. Y. Liao and Y. Yang, "Design and implementation of a mobile phone integrated alarming system," *2009 International Conference on Multimedia Information Networking and Security*, Hubei, 2009, pp. 65–68, Doi: 10.1109/MINES.2009.9.
9. H. Baqeel and S. Saeed, "Face detection authentication on smartphones: End users usability assessment experiences," *2019 International Conference on Computer and Information Sciences (ICCIS)*, Sakaka, Saudi Arabia, 2019, pp. 1–6, Doi: 10.1109/ICCISci.2019.8716452.
10. M. Pathak and A. K. Jayanthy, "Designing of a single channel EEG acquisition system for detection of drowsiness," *2017 International Conference on Wireless Communications, Signal Processing and Networking (WiSPNET)*, Chennai, 2017, pp. 1364–1368, Doi: 10.1109/WiSPNET.2017.8299986.

11. R. Inoue, T. Sugi, Y. Matsuda, S. Goto, H. Nohira and R. Mase, "Recording and characterization of EEGs by using wearable EEG device," *2019 19th International Conference on Control, Automation and Systems (ICCAS)*, Jeju, Korea (South), 2019, pp. 194–197, Doi: 10.23919/ICCAS47443.2019.8971564.

12. A. Badiei, S. Meshgini and A. Farzamnia, "Sleep arousal events detection using PNN-GBMO classifier based on EEG and ECG signals: A hybrid-learning model," *2020 28th Iranian Conference on Electrical Engineering (ICEE)*, Tabriz, Iran, 2020, pp. 1–6, Doi: 10.1109/ICEE50131.2020.9260671.

13. N. Leal, S. Moreno and E. Zurek, "Simple method for detecting visual saliencies based on dictionary learning and sparse coding," *2019 14th Iberian Conference on Information Systems and Technologies (CISTI)*, Coimbra, Portugal, 2019, pp. 1–5, Doi: 10.23919/CISTI.2019.8760988.

14. D. Mashru and V. Gandhi, "Detection of a drowsy state of the driver on road using wearable sensors: A survey," *2018 Second International Conference on Inventive Communication and Computational Technologies (ICICCT)*, Coimbatore, 2018, pp. 691–695, Doi: 10.1109/ICICCT.2018.8473245.

15. M. Choi, G. Koo, M. Seo and S. W. Kim, "Wearable device-based system to monitor a driver's stress, fatigue, and drowsiness," *IEEE Transactions on Instrumentation and Measurement*, vol. 67, no. 3, pp. 634–645, March 2018, Doi: 10.1109/TIM.2017.2779329.

16. I. Lopashchenko and O. Mazurkiewicz, "Estimation of face detection accuracy of Viola-Jones algorithm in video applications," *2018 International Conference on Information and Telecommunication Technologies and Radio Electronics (UkrMiCo)*, Odessa, Ukraine, 2018, pp. 1–4, Doi: 10.1109/UkrMiCo43733.2018.9047552.

17. X. Lian, "A face recognition approach based on computer vision," *2020 5th Inter-national Conference on Smart Grid and Electrical Automation (ICSGEA)*, Zhangjiajie, Chi-na, 2020, pp. 171–175, Doi: 10.1109/ICSGEA51094.2020.00044.

18. I. Lopashchenko and O. Mazurkiewicz, "Estimation of face detection accuracy of Viola-Jones algorithm in video applications," *2018 International Conference on Information and Telecommunication Technologies and Radio Electronics (UkrMiCo)*, Odessa, Ukraine, 2018, pp. 1–4, Doi: 10.1109/UkrMiCo43733.2018.9047552.

19. L. Tianhuang, D. Jian and Y. Wankou, "A multi-channel projection haar features method for face detection," *2017 32nd Youth Academic Annual Conference of Chinese Association of Automation (YAC)*, Hefei, 2017, pp. 148–153, Doi: 10.1109/YAC.2017.7967395.

20. Z. Yong, L. Jianyang, L. Hui and G. Xuehui, "Fatigue driving detection with modified ada-boost and fuzzy algorithm," *2018 Chinese Control And Decision Conference (CCDC)*, Shenyang, 2018, pp. 5971–5974, Doi: 10.1109/CCDC.2018.8408177.

21. S. Naido and R. R. Porle, "Face detection using colour and Haar features for indoor surveillance," *2020 IEEE 2nd International Conference on Artificial Intelligence in Engineering and Technology (IICAIET)*, Kota Kinabalu, Malaysia, 2020, pp. 1–5, Doi: 10.1109/IICAIET49801.2020.9257813.

22. P. Goel and S. Agarwal, "Hybrid approach of Haar Cascade classifiers and geometrical properties of facial features applied to illumination invariant gender classification system," *2018 International Conference on Computing Sciences*, Phagwara, 2018, pp. 132–136, Doi: 10.1109/ICCS.2018.40.

23. S. Emami and V. Suciu, (2018). "Facial recognition using OpenCV," *Journal of Mobile, Embedded and Distributed Systems*. vol. 4, pp. 38–43.

24. Q. Zhuang, Z. Kehua, J. Wang and Q. Chen, "Driver fatigue detection method based on eye states with pupil and iris segmentation," *IEEE Access*, vol. 8, pp. 173440–173449, 2020, Doi: 10.1109/ACCESS.2020.3025818.

25. B. Peng and A. K. Gopalakrishnan, "A face detection framework based on deep cascaded full convolutional neural networks," *2019 IEEE 4th International 2Conference on Computer and Communication Systems (ICCCS)*, Singapore, 2019, pp. 47–51, Doi: 10.1109/CCOMS.2019.8821692.

26. P. Drakopoulos, G. A. Koulieris and K. Mania, "Front camera eye tracking for mobile VR," *2020 IEEE Conference on Virtual Reality and 3D User Interfaces Abstracts and Workshops (VRW)*, Atlanta, GA, USA, 2020, pp. 642–643, Doi: 10.1109/VRW50115.2020.00172.

27. A. Singh, M. Singh and B. Singh, "Face detection and eyes extraction using sobel edge detection and morphological operations," *2016 Conference on Advances in Signal Processing (CASP)*, Pune, 2016, pp. 295–300, Doi: 10.1109/CASP.2016.7746183.

28. A. A. Joy and M. A. M. Hasan, "A hybrid approach of feature selection and feature extraction for hyperspectral image classification," *2019 International Conference on Computer, Communication, Chemical, Materials and Electronic Engineering (IC4ME2)*, Rajshahi, Bangladesh, 2019, pp. 1–4, Doi: 10.1109/IC4ME247184.2019.9036617.

29. L. Nahar, R. palit, A. Kafil, Z. Sultana and S. Akter, "Real time driver drowsiness monitoring system by eye tracking using mean shift algorithm," *2019 5th International Conference on Advances in Electrical Engineering (ICAEE)*, Dhaka, Bangladesh, 2019, pp. 27–31, Doi: 10.1109/ICAEE48663.2019.8975593.

30. S. Darshana, D. Fernando, S. Jayawardena, S. Wickramanayake and C. DeSilva, "Efficient PERCLOS and gaze measurement methodologies to estimate driver attention in real time," *2014 5th International Conference on Intelligent Systems, Modelling and Simulation*, Langkawi, 2014, pp. 289–294, Doi: 10.1109/ISMS.2014.56.

Chapter 14

An efficient image encryption scheme combining Rubik's cube principle with masking

M. Palav, and S.V.S.S.N.V.G Krishna Murthy
D.I.A.T

R. Telang Gode
National Defence Academy

CONTENTS

14.1 INTRODUCTION

With the advent of technical and digital revolution in the 21st century, a huge amount of information is disseminated over the Internet as pictures. Consequently, the drawbacks of unauthorized copying as well as unlawful sharing of digital multimedia contents also came into picture. With the speedy

DOI: 10.1201/9781003303053-16

expansion of computer-based telecommunications and network technology, the safety and privacy of image transmission have emerged as an important facet in the context of secure multimedia communications. In particular, the communications related to the important armed forces projects and defense applications have emerged as a prominent reason for the protected and secured transmission of digital contents. To overcome this challenge, researchers are trying to develop multimedia protection techniques that are robust and effective. In this circumstance, techniques such as encryption, steganography, and digital watermarking come into picture. Encryption of data is the process of changing data into an incomprehensible form using an algorithm that makes it indecipherable to anyone except for the authorized person. Conventional private key enciphering standards, such as DES, Triple DES, and AES, and public key enciphering techniques such as ECC and RSA may be unsuitable for image enciphering, especially for data related to real-time transfer because of some inhibit features of images like bulk data capacities, high correlations among pixel, and more redundancy. In the last 10 years, considerable research has been done on designing efficient encryption algorithms for secure transmission of real-time image data. In the literature, most of the encryption algorithms are based on permutation of pixels; these had already been found vulnerable to the known plaintext attack as well as chosen plaintext attacks and cipher text only attack. The high information redundancy leads to the breaking of secret permutations by comparing the permuted ciphertexts and the plaintext. The chaos-based image enciphering schemes [1–7] are generally considered extra trustworthy, but the cost of computation increases and especially one-dimensional chaotic map is restricted by its confined key space. The intrinsic features of chaos-based scheme are vulnerability to starting condition and randomness in behavior, etc. Image can be categorized as gray scale and color. A gray scale type image is of only one color and its pixel values range from [0–255], whereas a color picture is three-band monochrome image data. For example, RGB images represent red, green, and blue bands on separation of their components. The speed factor is of great concern when a large amount of high-resolution color image information needs to be processed. In 2013, the authors of [8] developed an enciphering scheme for colored image by only rearranging the pixel values of RGB. A chaos centered heterogeneous color image encryption based on permutation of bits was proposed by Wang [9]. One may refer to [10,11] for some multiple and Chebyshev sine chaotic schemes.

Keeping in view the efficiency with respect to speed as well as security, we aim to design an efficient image encryption algorithm centered on the combination of principle of Rubik's cube with chaos-based masking method. Image masking [12,13] is another interesting area of research that can combine with encryption techniques to enhance the security of image transmission over public channel. In [14], a Rubik's cube related image enciphering algorithm was proposed. Our scheme is an integration of the principle of Rubik's cube with masking constructed on symmetric tent chaotic map for optimizing the speed and security performance. First, we use the Rubik's cube encryption algorithm to permute the pixels of original image of size using two random secret vector keys of length m and n, respectively. The Rubik's cube encryption is explained in Section 14.3. For enriching security features of the projected algorithm, keys of the Rubik's cube method are used to generate key of asymmetric tent map, which further creates a 3×3 mask. Applying that mask to the permutated image, we get the final encrypted image.

Simulations as well as security testing have been carried out. The proposed algorithm comprises two main steps: first, the Rubik's cube encryption process for scrambling the pixels. Second, for achieving higher sensitivity and higher security, the algorithm further applies a 3×3 masking process.

Section 14.2 describes some preliminaries. The main Rubik's cube algorithm is explained in Section 14.3 followed by masking encryption in Section 14.3.4. Combined algorithms for complete encryption and decryption are given in Section 14.3.5. Experimental results along with security analysis are given in Section 14.4. Finally, the performance comparison of our proposed algorithm with recent relevant techniques is presented in Section 14.4.9.

14.2 PRELIMINARY SECTION

14.2.1 Chaotic maps

Chaotic maps [15,16] are generally used to create a sequence of random numbers that are mostly used for diffusing or scrambling the image. Since chaotic maps are very sensitive towards, control parameters and initial point (called key of chaotic map), they give a high encryption rate. In the scheme, we have used the one-dimensional asymmetric tent map to generate key for masking. The asymmetric tent map is defined as follows:

$$
x_n = \begin{cases} \dfrac{x_n}{\mu} & \text{if } 0 < x_n \leq \mu \\ \dfrac{1-x_n}{1-\mu} & \text{otherwise} \end{cases} \tag{14.1}
$$

Here, x_1 is the key of asymmetric tent map $x_1 \in (0,1)$ and μ is called the control parameter of the map and $\mu \in (0,1)$.

14.2.2 Masking

There are some operations in which the value of pixel value depends on the value of neighboring pixel value such subimages have the same dimension as the neighborhood. Such operations are called mask or filter. The value of mask is considered as coefficient. In this process, the mask moves from point to point in image at each point (x,y) and the corresponding value of mask is calculated by using the relation that is predefined. Mostly, masking method is used to improve image quality, but in our work, masking is used to encrypt image. The relation constructed is given by

$$q = \sum_{a=-1}^{1}\sum_{b=-1}^{1} w(a,b) \cdot f(x+a,y+b) \qquad (14.2)$$

Here, a,b takes integer values.

$$g(x,y) = (f(x,y)+q) \bmod 256 \qquad (14.3)$$

where $x = 0,1,...,m-1$, $y = 0,1,...,n-1$ and $f(x,y)$ and $g(x,y)$ are pixel data at position (x,y) of actual image and cipher image (here image is of size $m \times n$) and the mask is $w(x,y)$.

Next, we find the reverse or inverse of masking. Original masking is started from the upper left most corner, while in inverse masking, we begin from the lowermost right corner and, in this case, the variation is in the function which is given as $f(x,y) = g(x,y) - q_1$, where q_1 is obtained by using mask in reverse order on encrypted image and using equation (14.2), i.e., $q_1 = \sum_{a=-1}^{1}\sum_{b=-1}^{1} w(a,b) \cdot g(x+a,y+b)$. In this process of masking (or inverse masking) at each step, we apply a new pixel value to encipher (or decipher) the next value of pixel.

14.3 PROPOSED WORK

14.3.1 Rubik's cube encryption

The Rubik's cube encryption as explained in the following sections just permutes the position of the pixel value. Using two vector keys of length m and n, bitwise XOR operation is performed on rows as well as columns as described in the following section.

14.3.2 Rubik's cube encryption algorithm

The stages of encryption are given as follows:

Let $M = [M(i,j)]_{m \times n}$ be α-bit image of size $m \times n$. Let T_m and T_n be two random key vectors of length m and n, respectively. Components $T_m(i)$ for $i = 1,2,...,m$ of key vector T_m and components $T_n(j)$ for $j = 1,2,...,n$ of key vector T_n.

1. For each row i of image $\left[M(i,j) \right]$
   a. Calculate the sum of row i, that is denoted by $r(i)$ and given as

$$r(i) = \sum_{j=1}^{n} M(i,j) \qquad \text{for } i = 1,2,\ldots,m$$

   b. If $r(i)$ is even, then move each pixel position of row i to left and the first pixel moves to the last pixel.
   c. If $r(i)$ is odd, then move each pixel position of row i to right and the last pixel moves to the first pixel.
   d. Denote the obtained new image by M_1.
2. For each column $j = 1,2,\ldots,n$ of this new image M_1 obtained in Step 1
   a. Calculate the sum of column j, that is denoted by $c(i)$ and given as:

$$c(j) = \sum_{i=1}^{m} M_1(x,y) \qquad \text{for } j = 1,2,\ldots,n$$

   b. If $c(i)$ is even, then move each pixel position of column j to upward and the first pixel moves to the last pixel.
   c. If $c(i)$ is odd, then move each pixel position of row i to downward and the last pixel moves to the first one.
   d. Denote the obtained new image by M_2.
3. Using vector key T_m, bitwise XOR is operated on each column j of the scrambled matrix M_2 obtained in Step 2 as follows:
   For each column j:
   a. If i is even, then $M_3(i,j) = T_m(i)$ XOR $M_2(i,j)$.
   b. If i is odd, then $M_3(i,j) = T_m'(i)$ XOR $M_2(i,j)$.

   Here, vector $T_m'(i)$ is the vector obtained from $T_m(i)$ by circular shifting each component of $T_m(i)$ to left (the first pixel moves to the last pixel).
4. Using vector key T_n, bitwise XOR operation is applied on each row i of the scrambled matrix M_3 obtained in Step 3 as follows:
   For each row i:
   a. If j is even, then $M_4(i,j) = T_n(j)$ XOR $M_3(i,j)$.
   b. If j is odd, then $M_4(i,j) = T_n'(j)$ XOR $M_3(i,j)$.

Here, vector $T_n'(j)$ is the vector obtained from $T_n(j)$ by circular shifting each component of $T_n(j)$ to left (the first pixel moves to the last pixel).

M_4 is the final encrypted image (image in which the pixel of original matrix M is permuted) obtained by using the Rubik's cube encryption algorithm.

14.3.3 Rubik's cube decryption algorithm

Let D be the encrypted image obtained by Rubik's cube encryption algorithm. Steps of decryption are as follows:

1. Using vector key T_n, bitwise XOR operation is applied on each row i of the matrix D obtained as follows:

    For each row i:

    a. If j is even, then $D_1(i,j) = T_n(j)$ XOR $D(i,j)$.

    b. If j is odd, then $D_1(i,j) = T'_n(j)$ XOR $D(i,j)$.

    Here, vector $T'_n(j)$ is the vector obtained from $T_n(j)$ by circular shifting each component of $T_n(j)$ to left (the first pixel moves to the last pixel).

2. Using vector key T_m, on each column j of the scrambled matrix D_1 obtained in Step 1, bitwise XOR operation is performed as follows:

    For each column j:

    a. If i is even, then $D_2(i,j) = T_m(i)$ XOR $D_1(i,j)$.

    b. If i is odd, then $D_2(i,j) = T'_m(i)$ XOR $D_1(i,j)$.

    Here, vector $T'_m(i)$ is the vector obtained from $T_m(i)$ by circular shifting each component of $T_m(i)$ to left (the first pixel moves to the last pixel).

3. For each column $j = 1,2,...n$ of this new image D_2 obtained in Step 2

    a. Calculate the sum of column j, that is denoted by $c(j)$ and given as

$$c(j) = \sum_{i=1}^{m} D_2(x,y) \qquad \text{for } j = 1,2,...,n$$

    b. If $c(j)$ is even, then move each pixel position of column j to upward and the first pixel moves to the last pixel.

    c. If $c(j)$ is odd, then move each pixel position of row i to downward and the last pixel moves to the first pixel.

    d. Denote the obtained new image by D_3.

4. Next for each row i of D_3

    a. Calculate the sum of row i, that is denoted by $r(i)$ and given as

$$r(i) = \sum_{j=1}^{n} M(i,j) \qquad \text{for } i = 1,2,...,m$$

    b. If $r(i)$ is even, then move each pixel position of row i to left and the first pixel moves to the last pixel.

    c. If $r(i)$ is odd, then move each pixel position of row i to right and the last pixel moves to the first pixel.

    d. Finally, we get the final decrypted image.

14.3.4 Encryption using masking

Key Generation of Chaotic Map

Calculate $Q = \sum_{i=1}^{m} T_m(i)$ and $Q_1 = \sum_{i=1}^{n} T_n(i)$ and $h = Q + Q_1$. Then

$$\text{Key of chaotic map} = \frac{h}{10^{k+1}} \tag{14.3}$$

where k is the number of digits of h.

Generation of Mask

Using the asymmetric tent map with initial point (key of the chaotic map), generate a sequence of 9 numbers $\{l_1, l_2, ..., l_9\}$. l_1 is obtained from equation (14.3). Then transform the elements of the sequence $\{l_1, l_2, ..., l_9\}$ in range 0 to 255 using the following transformation map:

$$P_i = (l_i + c) \times 255 \quad \text{for} \quad i = 1, 2, ..., 9$$

$$\text{where} = \begin{cases} 0 & \text{for } i = 1 \\ (l_i \times 255) \bmod 1 & \text{for } i = 2, 3, ..., 9 \end{cases}.$$

Thus, the obtained 3×3 mask is $\begin{bmatrix} P_1 & P_2 & P_3 \\ P_4 & P_5 & P_6 \\ P_7 & P_8 & P_9 \end{bmatrix}$. Make center 0, i.e., $P_5 = 0$.

14.3.5 Proposed algorithm

Suppose M is the matrix of the original image. (Apply zero pedding to original image for masking.)

Encryption

Step 1: Apply Rubik's cube encryption algorithm on M and obtain encrypted image M_4.

Step 2: Using key vectors of Rubik's cube method and asymmetric tent map, create 3×3 mask.

Step 3: Apply mask to encrypted image M_4 to get the final encrypted image.

Decryption

Step 1: Using key vectors of Rubik's cube method and asymmetric tent map, create 3×3 mask.

Step 2: Apply reverse masking to the encrypted image to get D_4 the first decrypted image.

Step 3: Then perform Rubik's cube decryption algorithm and get the original image.

The flowcharts of encryption and decryption processes for grayscale and color images are given in Figures 14.1–14.4.

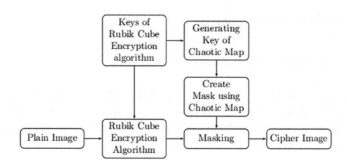

Figure 14.1 Flowchart: enciphering algorithm.

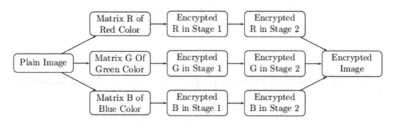

Figure 14.2 Flowchart: enciphering algorithm of color image, where Stage 1 is encryption using Rubik's cube principle and Stage 2 is encryption using masking.

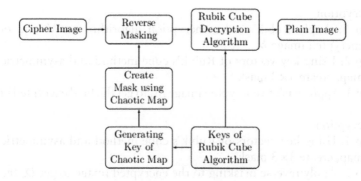

Figure 14.3 Flowchart: proposed decryption algorithm.

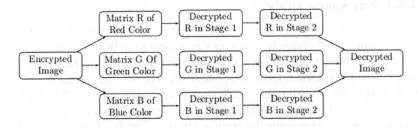

Figure 14.4 Flowchart: proposed enciphering algorithm of color image, where Stage 1 is reverse masking process and Stage 2 is decryption using Rubik's cube principle.

14.4 EXPERIMENTAL SETUP AND SIMULATION ANALYSIS

This process is applied on 256×256 as well as on 512×512 gray scale and color images. A system with configuration "64-bit, Intel® core™ i5-8250U 1.60GHz" has been used. The programming is done in MATLAB. Refer Figure 14.5.

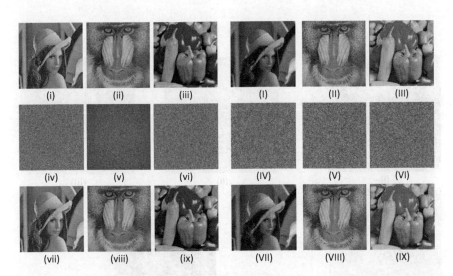

Figure 14.5 (i) Actual Lena image, (ii) actual baboon image, (iii) actual peppers image, (iv) enciphered image of (i), (v) enciphered image of (ii), (vi) enciphered image of (iii), (vii) decrypted image of (iv), (viii) decrypted image of (v), (ix) decrypted image of (vi), (I) colored Lena image, (II) colored baboon image, (III) colored peppers image, (IV) cipher image of (I), (V) cipher image of (II), (VI) cipher image of (III), (VII) decrypted image of (IV), (VIII) decrypted image of (V), (IX) decrypted image of (VI).

14.4.1 Key space analysis

Encryption scheme is safe from attack based on brute force if key space is greater than 2^{100}. If image is of size $m \times n$ in α-bit gray 0000000scale, then according to the proposed algorithm, key vectors T_m and T_n can take $2^{\alpha m}$ and $2^{\alpha n}$ possible values, respectively. Therefore, the total key space is $2^{\alpha(m+n)}$. For instance, for an 8-bit gray scale type image of size 512×512, the key space is 2^{8192}, which is greater than 2^{100} which makes this algorithm secure from brute force. (For eight-bit gray scale type image of size $m \times n$, this algorithm is safe if $m, n \geq 3$.)

14.4.2 Analysis of key sensitivity

The algorithm is key sensitive; if there is a minor change in key, the encrypted image changes completely and also the actual image is not easy to recovered if there is a slight change in the decryption key. Our scheme is key sensitive as depicted in Figure 14.6.

14.4.3 Histogram analysis

An encryption scheme is considered as good if pixel distribution of the encrypted image is uniform. Since the histogram of cipher image obtained

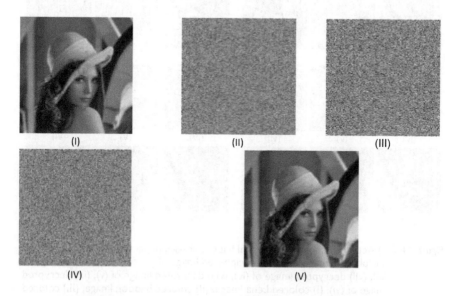

Figure 14.6 (I) Actual Lena image, (II) image enciphered using correct key, (III) image (II) decrypted using wrong key, (IV) image (II) decrypted using the key that is only one bit different from the correct key of Rubik's cube encryption, (V) image (II) decrypted using the correct key.

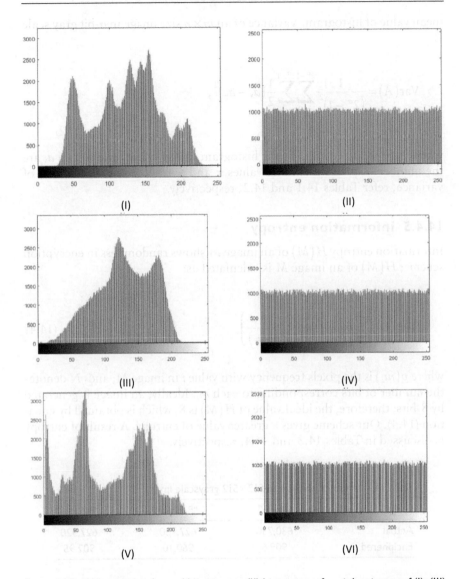

Figure 14.7 (I)Histogram of actual Lena image, (II) histogram of encipher image of (I), (III) histogram actual baboon image, (IV) histogram of encipher image of (III), (V) histogram of actual pepper image, (VI) histogram of encipher image of (V).

from the proposed scheme is uniformly distributed, therefore the scheme is safe from statistical attacks. Refer Figure 14.7 for details.

14.4.4 Variance

To analyze the uniformness of histogram of image quantitatively, we measure variance. Variance measures deviation between histogram and the mean value of histogram. Variance of an $m \times n$ size image in α-bit gray scale is

$$\text{Var}(A) = \frac{1}{\left(2^\alpha - 1\right)^2} \sum_{x=0}^{2^\alpha-1} \sum_{y=0}^{2^\alpha-1} \frac{1}{2}\left(a_x - a_y\right)^2,$$

where $A = \left[a_0, a_1, \ldots, a_{2^\alpha-1}\right]$ is the histogram values vector, and a_x and a_y are the number of pixels assuming values x and y respectively. For results of variance, refer Tables 14.1 and 14.2, respectively.

14.4.5 Information entropy

Information entropy $H(M)$ of an image M shows randomness in encryption scheme; $H(M)$ of an image M is calculated as:

$$H(M) = \sum_{i=0}^{2^N-1} p(m_i) \cdot \log_2\left(\frac{1}{p(m_i)}\right) \tag{14.4}$$

where $p(m_i)$ is the pixels frequency with value i in image M, and N denotes the number of bits corresponding to each m_i. Ideally, an image is generated by 8-bits; therefore, the ideal value of $H(M)$ is 8, which is obtained by equation (14.4). Our scheme gives a greater value of entropy. A result of entropy is discussed in Tables 14.3 and 14.4, respectively.

Table 14.1 Variance analysis of 512×512 grayscale image

Image	"Lena"	"Baboon"	"Peppers"
Actual	630,730	627,730	627,520
Enciphered	909.6	960.10	902.95

Table 14.2 Variance analysis of 256×256 color image

| Image | Red | | Green | | Blue | |
	Actual	Enciphered	Actual	Enciphered	Actual	Enciphered
"Lena"	27,154	257.42	57,735	244.8438	140,570	245.15
"Baboon"	17,939	266.13	363,940	259.60	190,910	221.30
"Peppers"	53,024	281.34	88,272	231.2969	125,180	293.24

Table 14.3 Information entropy analysis of 512×512 grayscale image

Image	"Lena"	"Baboon"	"Peppers"
"Actual"	7.3871	7.3579	7.5715
"Enciphered"	7.9994	7.9994	7.9994

Table 14.4 Information entropy analysis of 256×256 color image

| Image | Red | | Green | | Blue | |
	"Actual"	"Enciphered"	"Actual"	"Enciphered"	"Actual"	"Enciphered"
"Lena"	7.6729	7.9972	7.4087	7.9973	6.9635	7.9973
"Baboon"	7.7586	7.9971	7.4685	7.9971	7.7709	7.9976
"Peppers"	7.3402	7.9969	7.4770	7.9975	7.0569	7.9968

14.4.6 Correlation coefficient

In an image, an arbitrarily selected pixel is strongly related to the pixels in its neighborhood. The encryption scheme should be such that the cipher image has the least correlation coefficient. The scheme is good if the correlation coefficient of the enciphered image is close to zero. The analysis of correlation coefficient is shown in Table 14.5 and Figure 14.8.

14.4.7 Differential attack

Resistance against differential attack is measured using two terms: "Number of Pixel Change Rate (NPCR)" and "Unified Average Change Intensity (UACI)". The UACI and NPCR values for an image of size $m \times n$ are calculated as given below.

$$\text{NPCR} = \frac{\sum_{i=1}^{m}\sum_{j=1}^{n} R(i,j)}{m \times n} \times 100$$

Table 14.5 Correlation coefficient analysis of 512×512 grayscale image

| Image | Coefficient of correlation | | | | | |
| | "Horizontal" | | "Vertical" | | "Diagonal" | |
	"Actual"	"Enciphered"	"Actual"	"Enciphered"	"Actual"	"Enciphered"
"Lena"	0.9722	0.0017	0.9853	0.000836	0.9593	0.000172
"Baboon"	0.8665	−0.0014	0.8558	−0.0007	0.7263	0.0021
"Peppers"	0.9807	−0.00039	0.9752	−0.00082	0.9636	−0.00025

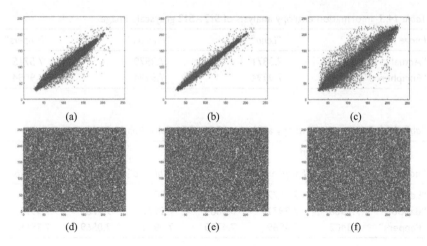

Figure 14.8 (a) Horizontal correlation plot of actual Lena image, (b) vertical correlation plot of actual Lena image, (c) diagonal correlation plot of actual Lena image, (d) horizontal correlation plot of enciphered Lena image, (e) vertical correlation plot of ciphered Lena image, (f) diagonal correlation plot of ciphered Lena image.

and

$$\text{UACI} = \frac{1}{m \times n} \left[\sum_{i=1}^{m} \sum_{j=1}^{n} \frac{\left| K_1(i,j) - K_2(i,j) \right|}{2^{\alpha} - 1} \right] \times 100$$

where

$$R(i,j) = \begin{cases} 0 & \text{if } K_1(i,j) = K_2(i,j) \\ 1 & \text{if } K_1(i,j) \neq K_2(i,j) \end{cases}$$

Here, K_1 and K_2 are enciphered images obtained by two actual images (size $m \times n$) with the one pixel difference. The results are given in Tables 14.6 and 14.7, respectively.

Table 14.6 UACI and NPCR analysis of 512×512 grayscale image

Image	"Lena"	"Baboon"	"Peppers"
NPCR	99.6222	99.6249	99.6226
UACI	33.4842	33.5237	33.5157

Table 14.7 UACI and NPCR analysis of 256×256 color image

Image	NPCR			UACI		
	Red	Green	Blue	Red	Green	Blue
"Lena"	99.6444	99.6434	99.6436	33.5489	33.5547	33.5490
"Baboon"	99.6400	99.6420	99.6399	33.5500	33.5589	33.5464
"Peppers"	99.6398	99.6400	99.6413	33.5633	33.5738	33.5591

14.4.8 Time analysis

Encryption speed evaluates the real-life potential use of algorithm. Any encryption scheme is considered to be good if it provides more security at a faster speed. Time distribution for encrypting different size images are depicted in Tables 14.8 and 14.9.

14.4.9 Performance comparison

Our scheme is compared with the recently developed schemes in terms of various performance factors as shown in Tables 14.10–14.12. This comparison shows the proposed scheme provides better results compared to other schemes. Also, it takes less time than other scheme.

Table 14.8 Time distribution of 512×512 grayscale image

Process	Rubik's cube encryption	Masking	Other	Total
Time (Sec.)	0.181	0.573	0.008	0.762
Percentage (%)	23.7532	75.1968	1.05	100

Table 14.9 Time analysis

Image	Grayscale image		Color image	
	Encryption time	Decryption time	Encryption time	Decryption time
128×128	0.063	0.064	0.161	0.162
256×256	0.209	0.212	0.576	0.584
304×512	0.465	0.473	1.338	1.347
512×512	0.762	0.784	2.221	2.231

Table 14.10 Comparison: implementation of some contemporary encryption
techniques on gray scale "Lena 512×512 image"

Factor of performance		Luo [17]	Dewahdeh [18]	Kheled [14]	Proposed algorithm
Image size		Square	Non-square and square	Square	Non-square and square
Correlation coefficient	Horizontal	0.0019	Not given	0.0068	0.0017
	Vertical	−0.0024	Not given	0.0091	0.000836
	Diagonal	0.0011	Not given	0.0063	0.000172
Differential attack	NPCR	99.6113	Not given	99.6111	99.6222
	UACI	33.4682	30.4814	28.621	33.4842
Sensitivity to keys		YES	YES	YES	YES
Variance		980.8	Not given	Not given	909.5
Entropy		7.9993	7.9970	7.9968	7.9995

Table 14.11 Time comparison with recent encryption schemes – grayscale image

Image size	256×256	512×512
Proposed algorithm	0.209	0.762
Norouzi [19]	0.31	–
Luo [17]	1.170844	4.73389
Chen [20]	5.556799	8.9744393
Dawahdeh Z.E. [18]	1.2615	–
Rehman A.U. [21]	7.73	31.59

Table 14.12 Time comparison with recent encryption schemes – color image

Image size	Proposed algorithm	Butt K [22]	Sharma A [23]	Seyedzadeh S. [24]
256×256	0.576	2.09	56.911	–
512×512	2.221	3.77	60.641	41.1227

14.5 CONCLUSIONS

This paper presented a novel, efficient image enciphering scheme centered
around the structure of Rubik's cube permutating image pixels followed by
masking to strengthen the security. The results of simulations with an exten-
sive security analysis depict the sturdiness of scheme against various statistical
as well as differential attacks. The algorithm shows better results with respect

to speed and is capable of fast encryption. The algorithm works on all sizes of gray scale image and is further extended to color images by implementing it on each of the color components. The proposed scheme is an efficient candidate and potentially useful for real-time image communication in a secure way.

REFERENCES

1. Abitha, K., Bharathan, P., 2016. Secure communication based on Rubik's cube algorithm and chaotic baker map. *Procedia Technology* 24, 782–789. Doi: 10.1016/j.protcy.2016.05.089.
2. Arpacı, B., Kurt, E., Celik, K., 2019. A new algorithm for the colored image encryption via the modified Chua's circuit. *Engineering Science and Technology, An International Journal*. Doi: 10.1016/j.jestch.2019.09.001.
3. Baptista, M.d., 1998. Cryptography with chaos. *Physics Letters A* 240, 50–54. Doi: 10.1016/S0375-9601(98)00086-3.
4. Chai, X., Fu, X., Gan, Z., Lu, Y., Chen, Y., 2018. A color image cryptosystem based on dynamic DNA encryption and chaos. *Signal Processing* 155. Doi: 10.1016/j.sigpro.2018.09.029.
5. Matthews, R., 1989. On the derivation of a "chaotic" encryption algorithm. *Cryptologia* 8, 29–41. Doi: 10.1080/0161-118991863745.
6. Tang, Z., Yang, Y., Shijie, X., Yu, C., Zhang, X., 2019. Image encryption with double spiral scans and chaotic maps. *Security and Communication Networks* 2019, 1–15. Doi: 10.1155/2019/8694678.
7. Yang, F., Mou, J., Sun, K., Cao, Y., Jin, J., 2019. Color image compression encryption algorithm based on fractional-order memristor chaotic circuit. *IEEE Access* Doi: 10.1109/ACCESS.2019.2914722.
8. Kester, Q.A., Nana, L., Pascu, A., 2013. A novel cryptographic encryption technique for securing digital images in the cloud using AES and RGB pixel displacement. Doi:10.1109/EMS.2013.51.
9. Wang, X., Zhang, H.l., 2015. A color image encryption with heterogeneous bit-permutation and correlated chaos. Optics Communications 342. Doi: 10.1016/j.optcom.2014.12.043.
10. Liu, H., Wen, F., Kadir, A., 2018a. Construction of a new 2d Chebyshev-sine map and its application to color image encryption. *Multimedia Tools and Applications* 78. Doi:10.1007/s11042-018-6996-z.
11. Som, S., Datta, S., Singha, R., Kotal, A., Palit, S., 2015. Confusion and diffusion of color images with multiple chaotic maps and chaosbased pseudorandom binary number generator. *Nonlinear Dynamics* 80. Doi: 10.1007/s11071-015-1893-8.
12. Cheikh, F.A., Khriji, L., Gabbouj, M., 1998. Unsharp masking-based approach for color image processing, in: *9th European Signal Processing Conference (EUSIPCO 1998)*, pp. 1–4.
13. Palav M., Gode R.T., Krishna Murthy S.V.S.S.N.V.G., 2021. A masking-based image encryption scheme using chaotic map and elliptic curve cryptography. In: *Mathematical Modeling, Computational Intelligence Techniques and Renewable Energy. Advances in Intelligent Systems and Computing*, Vol. 1287. Springer, Singapore. Doi: 10.1007/978-981-15-9953-8_12.

14. Loukhaoukha, K., Chouinard, J.Y., Abdellah, B., 2012. A secure image encryption algorithm based on Rubik's cube principle. *Journal of Electrical and Computer Engineering*. Doi: 10.1155/2012/173931.
15. Kumari, M., Gupta, S., Sardana, P., 2017. A survey of image encryption algorithms. *3D Research* 8. Doi:10.1007/s13319-017-0148-5.
16. Liu, Z., Xia, T., Wang, J., 2018b. Image encryption technology based on fractional two-dimensional triangle function combination discrete chaotic map coupled with Menezes-Vanstone elliptic curve cryptosystem. *Discrete Dynamics in Nature and Society* 2018, 1–24. Doi: 10.1155/2018/4585083.
17. Luo, Y., Ouyang, X., Liu, J., Cao, L., 2019. An image encryption method based on elliptic curve Elgamal encryption and chaotic systems. *IEEE Access*, 1–1. Doi: 10.1109/ACCESS.2019.2906052.
18. Dawahdeh, Z., Yaakob, S., Othman, R.R., 2017. A new image encryption technique combining elliptic curve cryptosystem with hill cipher. *Journal of King Saud University - Computer and Information Sciences* 30, 349–355. Doi: 10.1016/j.jksuci.2017.06.004.
19. Norouzi, B., Seyedzadeh, S.M., Mirzakuchaki, S., Mosavi, M., 2013. A novel image encryption based on row-column, masking and main diffusion processes with hyper chaos. *Multimedia Tools and Applications* 74. Doi: 10.1007/s11042-013-1699-y.
20. Chen, J., Zhang, Y., Qi, L., Fu, C., Xu, L., 2017. Exploiting chaos-based compressed sensing and cryptographic algorithm for image encryption and compression. *Optics & Laser Technology* 99, 238–248. Doi: 10.1016/j.optlastec.2017.09.008.
21. Rehman, A., Liao, X., Kulsoom, A., Abbas, S., 2014. Selective encryption for gray images based on chaos and DNA complementary rules. *Multimedia Tools and Applications* 74. Doi: 10.1007/s11042-013-1828-7.
22. Butt, K., Li, G., Khan, S., Manzoor, S., 2020. Fast and efficient image encryption algorithm based on modular addition and Spd. *Entropy* 22, 112. Doi: 10.3390/e22010112.
23. Sharma, A., Venkatadri, M., Kamlesh, G., 2019. Design and implementation of key generation algorithm for secure image. *International Journal of Recent Technology and Engineering (IJRTE)* 8(2).
24. Seyedzadeh, S.M., Norouzi, B., Mosavi, M., Mirzakuchaki, S., 2015. A novel color image encryption algorithm based on spatial permutation and quantum chaotic map. *Nonlinear Dynamics* 81. Doi: 10.1007/ s11071-015-2008-2.

Index

For Product Safety Concerns and Information please contact our
EU representative GPSR@taylorandfrancis.com Taylor & Francis
Verlag GmbH, Kaufingerstraße 24, 80331 München, Germany